Meta-Analysis in Environmental Economics

Economy & Environment

VOLUME 12

The titles published in this series are listed at the end of this volume.

Meta-Analysis in Environmental Economics

by

J. C. J. M. van den Bergh

Free University Amsterdam, Amsterdam, The Netherlands

K. J. Button

George Mason University, Fairfax, VA, U.S.A.

P. Nijkamp

Free University Amsterdam, Amsterdam, The Netherlands

and

G. C. Pepping

Free University Amsterdam, Amsterdam, The Netherlands

In association with:

S. R. Baaijens
S. Greco
S. M. Jongma
J. Kerr
B. Matarazzo
G. Munda

KLUWER ACADEMIC PUBLISHERS

DORDRECHT / BOSTON / LONDON

A C.I.P. Catalogue record for this book is available from the Library of Congress

ISBN 978-90-481-4862-2

Published by Kluwer Academic Publishers,
P.O. Box 17, 3300 AA Dordrecht, The Netherlands.

Sold and distributed in the U.S.A. and Canada
by Kluwer Academic Publishers,
101 Philip Drive, Norwell, MA 02061, U.S.A.

In all other countries, sold and distributed
by Kluwer Academic Publishers,
P.O. Box 322, 3300 AH Dordrecht, The Netherlands.

Printed on acid-free paper

TABLE OF CONTENTS

Preface

The present publication addresses the issue of comparative analysis of environmental economic policy studies at a meso level. It is based on the principles of meta-analysis, which means that studies for policy preparation are itself analyzed and evaluated from a more general and comparative perspective. It sets out to review various approaches to deal with such meta-analysis and presents various applications of meta-analysis to distinct fields of environmental economics and environmental policy analysis.

The study itself is the result of a project financed by the EU DG XII programme on ENVIRONMENT, Area III (contract number EV5V-CT94-0388), which deals with 'Research on Economic and Social Aspects of Environmental Issues' (1994-1996). The scientific research was undertaken by teams from three different universities in Europe, viz. the Free University of Amsterdam also acting as the coordinating institution, the University of Catania and the University of Loughborough (UK). The teams have tried to further develop and apply comparative analysis of meso environmental policy based on modern meta-analysis. The latter approach is able to synthesize results from separate but similar studies. Its aim is to enforce a more rigorous and systematic framework for evaluating case studies. The research in this publication has not only focussed attention on conventional meta-analytical (i.e. statistical) techniques, but also on advanced rough set analysis. The concept of meta-analysis has a considerable history in the natural sciences (e.g. medical research), but only recently it has begun to influence the social sciences (in particular, psychology). This book aims to demonstrate the methodological merits and empirical strengths of meta-analysis for environmental economics and policy analysis.

The strong research collaboration between the participating teams has to be recognized here as a noteworthy critical success factor, not only in the methodological but also in the applied part of this book. In addition to the main authors, various other colleagues have contributed to the contents of this book. Their names and specific contributions are listed on the following page. The actual writing of this book has taught the authors an important lesson on meta-analysis which is so beautifully expressed in an old Chines saying: "If you read, you will forget; if you see, you will remember; if you do, you will understand".

Amsterdam, Washington January, 1997

The authors

Acknowledgements

The authors wish to thank in particular the following persons for their contribution to this book: Roman Slowinski for the use of the software package Roughdas in Chapters 9, 10, 11 and 12 to perform the rough set analyses, Salvatore Greco for his kind assistance in the data analyses underlying the cases in the same four chapters and Benedetto Matarazzo for his clear methodological reflection on rough set analysis.

PART A

META-ANALYSIS AND ENVIRONMENTAL POLICY EVALUATION

CHAPTER 1 ENVIRONMENTAL POLICY EVALUATION

1.1 Aims and Scope

Environmental policy analysis is increasingly addressing quality of life issues at the regional and sector levels. The idea of meso level environmental policies and strategies is becoming established in North-America and many European Union (EU) countries. It, therefore, seems timely to systematically analyze such policies and strategies, and to evaluate their performance. Such an analysis could take a number of forms. The focus here is on the application of meta-analysis to the subject. This means analysis at a meta-level, where 'meta' refers to the fact that policy analysis is itself analyzed and evaluated from a more general perspective. This requires not only an inventory of existing and past experiences on the basis of a literature review, but also a systematic and rigorous comparison and evaluation of these experiences. In this book the basis of comparative policy assessment, the principles of meta-analysis and a wide spectrum of environmental case studies at a meso level are presented. The goal is the development of a set of summarizing and generalizing guidelines which can improve the preparation and implementation of environmental policies and strategies.

1.2 Assessment and Evaluation of Environmental Analyses at a Meso Level

Environmental policy preparation requires a series of steps that are often separately developed and applied. These include the following:

- description of environmental problems and systems, preparation of appropriate data, and the choice of relevant performance indicators;
- analyses of static, spatial or dynamic characteristics of systems behaviour, strategies or policies in an environmental context;
- impact assessment, a monetary judgement or a multicriteria evaluation of data obtained from the analyses carried out;
- policy formulation and choice of instruments.

In performing the last phase, a multi-disciplinary and multi-dimensional policy-strategic perspective on issues is needed. This perspective enables one to examine how information obtained in the different steps may be aggregated or condensed, compared, and evaluated against economic criteria such as efficiency, cost-effectiveness, equality of distribution, and overall effectiveness.

Developed countries have different traditions in policy preparation and strategic decision-making, which often differ in their sophistication. Inconsistent and *ad hoc* sets of planning tools are frequently used. There is, therefore, ample scope for introducing more rigorous frameworks and guidelines, and for applying systematic evaluation and correction procedures for the preparation of environmental policies and for strategic decision-making.

Improvement may be brought about by describing and comparing different approaches to environmental policy preparation and analysis at the meso level. This level of environmental policy and strategy making brings in a clear spatial dimension, embracing, for instance, transport externalities, tourism, energy and resources, agriculture, land use, and multisectoral industries. At present, rigorous comparison and analysis of such approaches are lacking. Evaluation has tended to be concentrated at the single project level instead of the meso level. There has been only limited application of the systematic meta-approach looking at the policy and strategies aimed at meso environmental matters.

This book seeks to develop an integrating methodology for applying meta-approaches to environmental economics and environmental policy analysis at the regional and sector level. A meta-approach or meta-analysis includes a systematic framework with operational criteria and clear determinants of methods needed in environmental policy or strategy preparation. It can use both quantitative statistical and qualitative methods for more rigorous evaluation of past efforts and prevailing experiences in policy and strategy evaluation. The term meta-analysis refers to the use of such quantitative and qualitative methods. A meta-approach to policy analysis can help to create order out of the chaos which often exists in comparing the findings of various studies on similar issues, and, thus, it can help synthesize results from separate but similar studies (Wolf, 1986). Its aim is to enforce a rigorous and systematic evaluation of separately performed studies. It goes further than a conventional review of studies, and can be characterized as reducing, or at least making transparent, subjective elements which usually enter policy studies. Although meta-analysis already has a long history in the natural sciences, the technique was first used by social scientists on a large scale in the United States in the early 1970s, when efforts were made to assess the New Frontier and Great Society programmes (Wachter, 1988; and Mann, 1990).

Meta-policy analysis might ultimately create a more uniform framework for the harmonisation of environmental policies, especially in countries which strive for more international coordination of policies. This is particularly relevant in the European Union or in the emerging Pan-European context, but is likely to extend to other emerging blocks in both the developed and the developing world. This is important in view of the current position where the tendency is for unsystematic efforts to arrive at quick decisions concerning environmental policies, but which shall have strategic, long term and structural economic implications. The formulation of lessons from previous experiences, therefore, may be useful. Finally, the relevance of the topic extends to other multi-dimensional, multi-disciplinary complex policy issues in Organisation for Economic Cooperation and Development (OECD), Eastern European, and developing countries.

Linking of methods on similar research issues has received little systematic treatment until now. It requires both a methodological level of conceptualizing, thereby deciding about criteria and methodology, and an empirical level of operationalizing. In order to take the second step, inputs of different expertise are necessary. It is clear that this requires an ambitious research effort. Some of the important issues will receive more attention in this book than others. It is hoped, however, that the present set of results will be useful in environmental economics research as well as in policy preparation.

1.3 Relevance of Meta-approaches in Environmental Economics

In environmental economics, meta-analysis seems to have a major potential for both developing a consensus on point estimates in economic valuation of environmental degradation or improvement, and for exploring factors that have influenced variations in point estimates among individual studies. Meta-approaches may be used to summarise over a collection of similar relationships and indicators, or to average, possibly using weights, over a collection of values obtained in similar studies. It is suited to this because the point estimates are, by definition, quantified and, hence, conventional statistical methods can usually be applied with relative ease.

Within these boundaries there is a wide range of environmental evaluation issues which seem amenable to the deployment of meta-analysis techniques. Some are areas where a number of studies have already been attempted, others are areas where there seems to be future potential. Six particularly promising issues are:

Evaluation of environmental costs. Here we are concerned mainly with point estimates of the monetary valuations of particular environmental costs. Work in this area has expanded considerably in recent years, both at the theoretical level, but perhaps more so with regard to empirical estimation procedures. The variety of techniques grounded in revealed preference theory has grown, while a range of innovative stated preference, or contingent valuation, methods have emerged (Johansson, 1987; Freeman, 1993). Meta-analysis can be used to look for indicators of central tendency in previous studies. The types of valuation categories to be considered could include:

- Noise nuisance problems (especially in terms of effects of noise on property values)
- Recreational amenity (e.g. travel costs method studies)
- Visual intrusion
- Traffic accident levels
- Atmospheric pollution at the local level (e.g. contingent valuation methods and waste production studies)
- Global warming issues (here impact studies could be looked at but there are some problems of compatibility)
- Watercourse pollution (lost production methods and 'clean-up' cost methods)
- Traffic congestion (not a strict environmental issue but often related to it)
- Crop damage
- Urban traffic congestion.

While these are relatively well established areas of study, there are also a number of emerging topics where meta-analysis also exhibits potential as a tool for synthesis. The case of response functions for dealing with the effects of pollution on health and materials is one example. The inversed recognition that economic and ecological factors should be integrated in studies using soil quality, irrigation and labour as inputs into production functions in resource based activities such as agriculture and fisheries is producing a series of parameters which could be synthesised. Work on materials product systems (Kandelaars *et al.*, 1995) and industrial metabolism (Ayres, 1989) also reveals a growing body of studies where estimations are not uniform.

Alternatively, synthesis approaches can be deployed to seek explanations of why studies offer differing results, e.g. by examining the spatial and temporal contexts in which the various studies which have been completed. In addition, and using the same basic framework of meta-analysis, the technique can be deployed to explore the differing evaluation results derived when adopting various different methods of valuation in policy assessment. For instance, it may be useful to explore whether some techniques systematically give higher valuations than others.

The forecasting of non-direct effects of economic activities on the environment. Forecasting the direct, physical non-environmental impacts of any policy is difficult, but these impacts are frequently the main determinant of the ultimate scale of the environmental damage done. For instance, environmental economists have expended considerable energies and resources trying to place a money value on factors such as traffic noise nuisance (Button, 1995) or changes in the probability of fatal accidents taking place. But from the overall social perspective, if the actual traffic forecasts are seriously incorrect then the accuracy of these evaluations becomes of secondary relevance. A 5 percent improvement in the valuation of noise is easily swamped by a 20 percent under-prediction of future traffic volumes. Meta-analysis, by synthesising a range of forecasts and seeking moderator variables, can provide a useful basis for improving the forecasts of the direct physical implications of decisions which may then be fed into the assessment process.

Assessment of the effectiveness of alternative environmental policy instruments. Approaches to environmental policy have gradually changed in recent years with an upsurge of interest in the use of economic instruments (such as taxes, polluter charges and tradeable permits) and a movement away from an almost total reliance on command-and-control regulatory measures. There is, also, a growing, if still small, body of individual retrospective case studies which have sought to examine how successful different policy strategies (e.g. fiscal policies, regulation and moral suasion) have been in the environmental area. Conducting these types of study takes a considerable amount of time and a case can be made for pulling together the information that is available.

While the nature of the output of these studies may mean that any meta-analysis will often require the adoption of soft modelling, for example, qualitative response models, rather than more conventional statistical analysis, the situation is that there is now both the background material becoming available for such work, while relevant techniques are available or being developed to conduct fruitful meta-analysis. It is also clearly a subject of considerable practical importance at a time when the traditional command-and-control approach to environmental policy is being supplemented by fiscal instruments. It also has relevance in the context of combining appropriate environmental protection policy with other policy areas, such as economic policy, land-use policy and infrastructure investment, when the latter are being initiated rather than being treated simply as a later add-on.

Exploration of the appropriate political level of intervention to contain environmental damage. While the number of case studies is still relatively small, there are now an increasing number of studies which have explained the success in adopting different policy options at the local, state and federal levels. From the perspective of effective policy management, meta-analysis may prove helpful in pointing to ways in which administrative structures could be beneficially improved. In particular, by bringing together in a systematic way the findings of these studies it should be possible not only to decide on the most effective level of government to tackle specific environmental problems but also to explore linkages between administrative levels to avoid potentially counteractive effects. As an example, a local and well defined policy to reduce river water pollution may be totally ineffective if a state or national policy allows dumping up-stream.

The political acceptability of alternative environmental instruments by decision-makers. Despite some recent developments, there seems to be a strong traditional bias in favour of using regulatory rather than fiscal tools to combat adverse environmental impacts. The literature looking into why this is so, and looking at the particular preferences of different social groupings, is growing. The approach adopted usually involves the use of questionnaire studies eliciting the views of groups such as academics, bureaucrats, businessmen, politicians and the general public. The body of works available in this field has, however, differing international origins, look at different sub-sets of factors, embrace diverse groupings and so on. There is scope for systematically bringing together this information and analysing it in order to more fully understand the socio-political forces underlying current thinking on environmental policies.

Spatial considerations can also be important in environmental policy making. There are various levels of multiplier implications associated with policy decisions (Nijkamp *et al.*, 1990). Different studies have tended to adopt alternative approaches to measuring these environmental policy effects and the extent to which they are traced through the economic and social structure. While not always quantifiable, the importance of making due allowance for these effects at different levels of aggregation justifies looking at the potential for some qualitative examination of what could be achieved using a quasi-meta analysis (i.e., semi-statistical procedures). An important question, for example, may concern the conditions whereby secondary effects cease to be of any real importance to the outcome of an environmental policy option.

The conclusion to be drawn from this list of possible environmental applications of meta-analysis is that there is potentially a vast range of interesting assessment opportunities at different levels of policy-making and for different environmental concerns. Clearly, this list is by no means definitive, but should merely be seen as illustrative. It perhaps should be regarded as reflecting the current concerns of environmental policy making and analysis, and in the longer term a multiplicity of new uses may emerge.

1.4 Organisation of the Volume

The structure of the book is based on a top-down approach and is subdivided into four parts.

Part A deals with the policy level at which environmental decision-making takes place. It focuses on environmental management and policy and provides the context for meta-policy analysis and specific application areas in which meta-analysis may be of use for environmental policies. It deals with the types of policy evaluation methods used, how integration of these methods takes place and the relationship between the type of environmental policy or strategy preparation and the specific combination of methods used in their analysis. Given this background, a framework is developed for a meta-level approach to environmental policy analysis. Part A consists of Chapters 2 to 4. Frameworks and relevant criteria are in these chapters provided at the level of decision-making in environmental management (Chapter 2) and at the methodological level of meta-analysis (Chapters 3 and 4).

The second part, Part B, consists of an inventory of existing approaches in the field of meta-analysis of policy-strategic preparation at a meso level. This is partly based on a broad literature review. It deals with the various types of meta-analysis and past applications in economics, and proceeds with a discussion of the methodological issues of these techniques. This second part consists of Chapters 5 to 9. In Chapter 5 an inventory of existing approaches of meta-analysis is presented as well as a review of past applications. Subsequently, promising meta-analytical techniques are described, which are used in the case studies discussed later in the book. Chapter 5 presents the traditional, statistical methods for carrying out meta-analysis. Chapters 6 to 8 offer information on the new method of rough set analysis coming from data-analysis and decision theory.

Part C, consisting of Chapters 10 to 16, shows applications of meta-analysis to various fields of environmental economics and policy at the meso level. These empirical applications are carried out on the basis of the criteria and the provisional frameworks developed earlier. A systematic evaluation in each case is followed by conclusions on relevance and importance of determining factors. This requires a multidisciplinary expertise.

The final part, Part D, is the development of guidelines for adapting existing and creating new meta-approaches. The aim is to improve upon existing research practices in various countries with the objective of preparing and supporting environmental policy and strategies at the meso level. The guidelines follow as elaborated conclusions of the work done in Part B and C of the book. They will be formulated by using the sets of determining factors and criteria, from the perspective of specific environmental policy and strategy relevant issues.

CHAPTER 2 META-APPROACHES TO ENVIRONMENTAL POLICY ASSESSMENT

2.1 Introduction

Concern with environmental quality is, and is likely to remain, a central component of socio-economic policy formulation and debate. The intensity of concern and the particular aspects of the environment which attract attention at any point of time appears to vary significantly. In part, this reflects the society's preference function, in part it reflects information levels, and in part it is related to perceived trade-offs which have to be made. Despite these shifts in emphasis, there is a need for policy-makers to develop strategies which meet evolving states of the world, and to provide a basis for longer term initiatives. Policy assessment procedures have become an important component of this strategic process.

This chapter sets out to explore the underlying nature of the issues involved and to develop a systematic scheme to make transparent the types of consideration which need to be incorporated in the assessment process. There are many dimensions in this process, however, and this chapter is primarily concerned with ways in which those involved in policy assessment can make better use of existing information and knowledge obtained through empirical investigations. It does not, therefore, concern itself directly with matters such as planning institutions, appraisal techniques or policy implementation and monitoring. The argument developed is that there are often weaknesses in both the information used at different stages of policy-making and in the way in which it is used. This may impede both the flexibility of decision-making and ultimately be wasteful in resource utilisation. Clearly, improving this information base may, potentially, yield a high social return.

Initially, this chapter looks at some of the particular features of environmental problems and the difficulties these may pose for policy development at the micro and meso levels. The issues surrounding environmental policy evaluation are considered, especially in terms of the more effective utilisation of existing information and insights gained from previous experiences. The usefulness of employing meta-analysis for synthesis of research outcomes is emphasized. The chapter explores the areas in which meta-analysis could provide new insights in the assessment process by extracting more information from previous work but, in doing this, it also highlights areas where further primary research would yield a significant return in policy analysis.

2.2 Environmental Problems and Policy

In recent years, the concept of sustainable development has gained widespread attention and a high degree of acceptance as a long-run goal of environmental policy. In addition to meeting sustainability objectives, however, there is the need to design environmental and socio-economic policies that are acceptable, efficient and effective (Böjo *et al.*, 1990; Opschoor, 1992; Van den Bergh, 1996).

As part of the on-going debates surrounding the operationalisation of the sustainable development concept, particularly at the sub-global level, the use of impact analysis either *ex ante* or *ex post* has provoked much interest in the public arena and among policy analysts (see Giaoutzi and Nijkamp, 1993). Despite the gradual introduction of more sophisticated environmental impact analyses in many countries and the efforts of bodies such as the World Bank to develop standard techniques, we still observe severe criticisms regarding the appraisal of policies. These criticisms are heightened by instances were considerable diversity emerges in the results generated by different studies of similar issues. This offers confirmation to critics that the overall reliability of such results is far from satisfactory. In any case, the popular perception of scientific studies appears to exhibit a wide variety of interpretations, views, findings and lessons.

One important reason for variations in the results obtained is the complex multidimensional nature of societal and environmental systems and interactions. These incorporate numerous feedback effects and are often hard to trace and still harder to quantify, evaluate and forecast (Nijkamp *et al.*, 1990). This also usually means there is a need for the adoption of simultaneous rather than recursive systems of analysis. This former type of framework is particularly difficult to develop, given the knowledge we have of interactive effects.

Environmental impacts are also usually unequally distributed among members of society so that the equity issue inevitably receives a prominent place in any environmental debate. Spatial externalities, for instance, can lead to quite disparate views regarding the implications of the same form and scale of any physical environmental impact. The frequent result is that, through a complex array of externalities, biased decisions are often taken, which can aggravate environmental problems (Pearce and Nash, 1989).

Finally, policy responses to environmental problems often tend to be extreme and frequently to be either almost inert or over-reactive. This is, in part, an information problem, but also stems from institutional mechanisms and, on occasions, can be the result of the bluntness of the policy instruments at hand. The outcome is that market failures are heavily influenced, and often compounded, by government response failures (OECD, 1992).

Conflicting viewpoints are, therefore, often at the heart of environmental policy evaluation and trade-offs. Such conflicts usually provoke ambiguous policy choices characterised by a diversity of measures to attain more or less similar goals. This may, for example, take the form of inter-actor conflicts such as those which occur between environmental groups and land developers; they may be spatially determined, reflecting different interests of various affected regions; or they may be inter-temporal and concern the interest of current generations against those of future generations.

To cope with these types of conflicts, several styles of public policy may be distinguished (Nijkamp and Blaas, 1994). First, there is ameliorative problem solving that focuses attention on avoiding and tackling frictions. Second, there is allocative trend-modifying that involves developing a future structure and reference pattern as a frame of reference for current planning possibilities and directions. This may provoke the need for strategic environmental polices of a long range nature.

Next, with exploitative opportunity-seeking, future bottlenecks and problems are identified in order to design alternatives and to select the most favourable pattern of change. This is a rational type of policy style which has gained some popularity in terms of policy scenario analysis. Finally, there is normative and goal-orientation which aims at designing spatial alternatives on the basis of a priori aims.

While information on the implications of alternative policies and strategies is not at the root of these conflicts, better information can assist in the reaching of compromises and the disregarding of less suitable alternatives. Environmental policies in a given sector, however, while frequently having some common elements and objectives, are never repeated experiments. There is, therefore, a need for drawing lessons from a multiplicity of related policies under varying circumstances. This is essentially the subject matter of the present study.

2.3 Impact Assessment and Environmental Policy Evaluation

In general, any environmental assessment problem is composed of two parts, namely impact analysis and policy analysis (Nijkamp and Blaas, 1994; Pearce and Nash, 1989). If the processes are combined with the diverse styles of public policy and the conflicting interests of the various parties involved, then the complexity of the environmental evaluation process becomes clear. Some indication of this is gleaned from the schema set out in Figure 2.1. Further, evaluation may be carried out with different levels of ambitions and with different degrees of precision. In general, it can be helpful to distinguish between: assessment of the contribution of individual policy measures towards solving a given policy problem; consideration of the appropriateness of the package of policy instruments in the light of changed economic circumstances or the contentions of economic theory; measurements of the range of benefits accruing from the package of relevant policy measures; and estimation of the costs and benefits of the policy and the cost-effectiveness of individual policy instruments.

In the past decade, numerous evaluation and assessment studies of environmental impacts have been carried out. The need for a systematic analysis of the impacts of environmental policies, or indeed any policy with significant environmental implications, has therefore increased. This is even more so because, first, many countries and sub-national regions are now going through a process of drastic socio-economic and technological restructuring which is impacting on their citizens' quality of life. This dynamic evolutionary pattern exhibits a clean break with the past which was, to a large extent, marked by stable dynamics. In the present circumstances, assessing the influence of alternative policies is much more complex. Our current analytical models do not, in general, adequately reflect the present, much more turbulent, dynamics of industrial economies. They are even less likely to be able to handle future changes.

Second, slowing or declining growth, coupled with cuts in public finance budgets in many regions, have called forth a need for a more careful analysis of the effectiveness and efficiency of public expenditures. Consequently, a systematic appraisal of social and economic programmes has become an important issue in environmental policy analysis. Policy must now be structured in the context of fiscal expediency as well as simple cost-effectiveness. For this reason a comparative analysis based on different impact studies is important to increase efficiency in the public sector, and in environmental policy as part of this. In this context, meta-analysis provides a useful analytical development for structuring environmental policy analysis. It makes for greater efficiency in the use of existing data and can be used to develop benchmarks from previous policies against which new initiatives may be assessed.

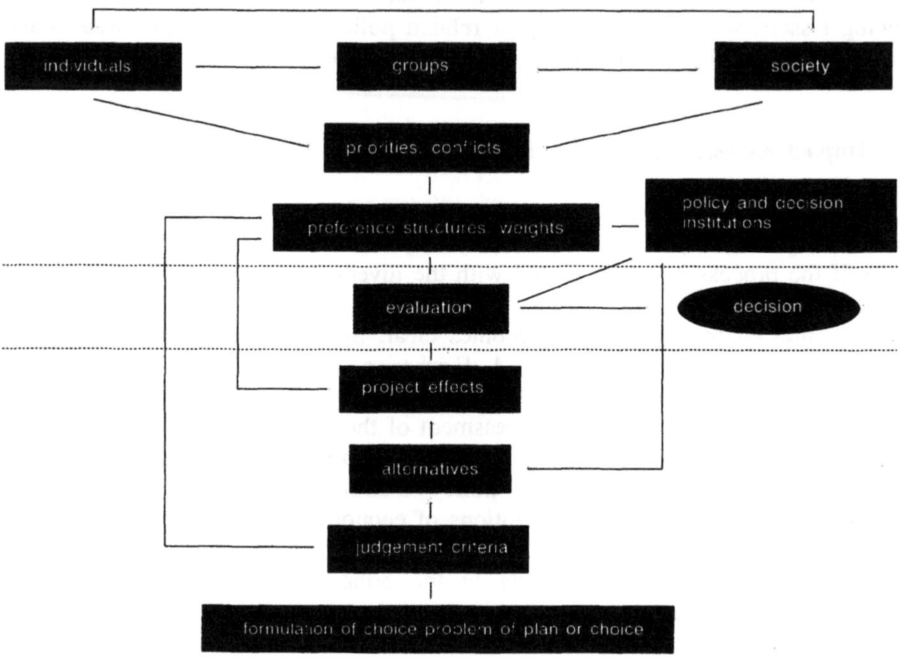

Figure 2.1. General structure of an evaluation problem.
Source: Nijkamp and Blaas (1994).

There are many areas in the field of impact assessment where serious methodological problems have emerged. For instance, should the type of impact to be assessed be treated equally for all kinds of instruments, groups or regions? How is one to isolate the effect of a certain policy instrument from other kinds of effects? The choice of an appropriate impact assessment method to be applied in a given situation is often sketchily dealt with in the literature.

There are two difficult aspects of impact analysis which deserve special attention, *viz.* the fuzzy nature of many policies related to imprecision in measurement and the uncertainty in defining operational indicators. While we pay more attention to the latter, some initial comments on fuzziness are justified.

The fuzzy nature of policy concerns both the policy objectives being sought and the effects of policy instruments used to attain these objectives (Munda *et al.*, 1993). It is widely recognised that effects upon a single decision criterion, as a result of different policy instruments, cannot always be measured in an unambiguous manner. This is because policy measures and goals may be of a quite vague or qualitative, and often multi-dimensional, nature. Instruments can, for instance, be subdivided into quantifiable instruments, qualitatively defined plans and broad legislative measures. Similarly, policy objectives may vary from quantifiable targets to qualitative policy desires.

Further, comparative assessments of values, objectives and instruments involve considerable uncertainty as the measures are generally based on a variety of different scales, ranging from cardinal scales to nominal or qualitative information. In some cases little more than verbal comment is possible. What is helpful is to be able to bring together the information which is available in such a way that additional work can be diverted where information is less precise. Traditional meta-analysis allows one to do this for factors which are quantified through the use of dummy variable regressions. However, for many other cases, recent methodological developments in qualitative or fuzzy data analysis, such as rough set analysis, now permit this narrow down process to go even further.

2.4 Policy, Contextual and Scientific Factors of Evaluation Methods

Prior to drawing up a framework for meta policy analysis, it is helpful to consider issues concerning the selection of descriptive, analytical and evaluation methods. The decision on method is seldom based entirely on technical considerations. In particular, irrespective of whether the motivation for a particular study is purely scientific or policy-oriented, management and policy characteristics ultimately play an important role in selection of method.

The logical starting point for gaining greater insights into specific characteristics is to distinguish between policy goals and instruments. To a large extent the choice of a specific method will be determined by the characteristics of the policy issue under consideration. For instance, the contingent valuation method, used to put monetary values on environmental externalities, is perhaps suited to estimating social costs resulting from local environmental problems, where people feel very much involved (Hoevenagel, 1994). Other methods are more germane for global matters such as greenhouse gas emissions.

Since the policy objective is generally the primary determinant of the method chosen, and derived dimensions are secondary, the following may constitute a set of criteria for deciding on method: (i) overall policy objectives; (ii) changes through which the policy objective may be realised; (iii) policy focus and constraints on behaviour; (iv) range of policy options and instruments; (v) focus and features of particular study; and (vi) assumptions on scientific relevance and system boundaries. The first four factors are at the level of policy, the fourth factor covers the contextual aspects, and the last factor is at a scientific level.

The objectives of environmental policy can be classified along the following lines. First, it may involve efforts to attain a more efficient allocation of resources, which closes the gap between social and private optima (a Pigouvian framework). Optimal allocation is couched in terms of the economic system, in time and in space. Second, there may be a distributional orientation, involving justice or, more narrowly, equity, within a generation or between generations or between regions or countries. Third, the objective may focus on effectiveness in realising a politically determined environmental quality level, a pollutant emission level or, more vaguely, environmental sustainability.

Attaining any of these objectives requires that feasibility and enforceability be guaranteed. These conditions were initially discussed in an economic policy setting by Tinbergen (1956) when developing arguments about links between the number of policy instruments and number of policy criteria. Further operational method developments and applications have taken place, treating the controlled or planned economy as a discrete (Theil, 1964) or continuous system (Chow, 1975).

Any environmental economic policy, when taken away from the abstract standard economic framework of social welfare optimisation, involves a trade-off between conflicting economic and environmental conservation interests. This gives rise to objectives such as social welfare optimisation, sustainable development, equity criteria related to inter- or intra-generational distribution, minimisation of environmental costs, or double dividend of eco-taxes. Simultaneously, policy formulation implies a trade-off between these objectives and those for which no objective social evaluation exists.

The implementation of these policy goals generally involves a diversity of change that can include:

- Changes in total volume of activities, e.g. in extraction, production and consumption.
- Changes in technology, e.g. increasing economic efficiency of existing activities; increasing environmental effectiveness or a shift to new process characteristics.
- Reallocation to activities, either of inputs (resource perspective) or of outputs (output perspective).
- Substitution between inputs (activity perspective) or between outputs (consumption perspectives).

The changes, necessary to achieve the policy goals, are normally realised via the behavioural responses of economic actors to the policy incentives - their behaviour being affected by a number of factors, some economic but others sociological.

Government environmental policy, at all geographical levels, deploys a variety of policy instruments, generally in packages, which are operationalised and administered through a range of institutions and agencies. These instruments may be categorised as:

- Direct regulation by a public authority, e.g. the introduction of physical standards or permits.
- Environmental legislation, involving property rights allocation, e.g. rights to environmental quality, and arrangement of liability for environmental damage.

- Price changes related to the use of economic or financial instruments, e.g. levies and taxes.
- Providing conditions for the introduction of new techniques, e.g. social R&D, and soft loans for environmental projects.
- Stimulating, organising and bargaining for voluntary agreements involving market participants, e.g. environmental covenants and eco-labelling of products.
- Moral suasion by conveying information to the public, e.g. through direct communication and education.

In addition to these instrument categories, policy options and instruments, other than those directly related to environmental policy, can impact on the policy targets. These influences can, for example, be through land use effects or nature policies, that can involve ecological infrastructure or restoration of ecosystems. They may also operate via economic policies such as industrial and sectoral subsidies and regulation, often involving transport and energy, or particular policies related to resource based activities such as agriculture, forestry, fisheries and mining. Finally, broader budgetary and finance policies, such as ecological taxation, may be relevant.

Table 2.1. Examples of characteristics of environmental-economics research.

Problems to be tackled	*Economic sectors*	*Spatial level*
pollution	industry	local
production	resource based	regional
consumption	agriculture	fluvial
specific pollutants	forestry	continental
organic	fishery	global
metals	recreation	*Technology*
reactive	transport	invention
toxic	services	innovation
interactive pollutants	public sector	materials efficiency
exhaustion of resources	*Processes*	end-of-pipe
non-renewable	extraction	production
renewable	abatement	recycling
energy	treatment	*Scientific aim*
land	product recycling	description
water	materials recycling	analysis
other problems	incineration	explaining
congestion	dumping	decomposition
noise	accumulation	testing
disturbance	decay	forecasting
cutting up	erosion	evaluation
	weathering	ranking

Once a case study is chosen to be included in a meta-analysis, there are a number of specific characteristics which are part of, but not explicit to, the policy objectives, desired changes or instruments. They might, however, be important in making decisions concerning the types of policies to apply. Table 2.1 offers an indication of the possible range of such factors, taking account of spatial aggregation, time schedules, specific focuses, problems, activities, processes and technologies.

The final category of factors important in meta-analysis method selection stems from the fact that every study makes explicit and implicit assumptions regarding which variables to include, and how to classify them as endogenous variables and exogenous variables. Choices impact on the appropriate method. They determine how general the analysis will ultimately be. While it is impossible to be categorical, Table 2.2 offers some broad generalisations concerning the instrument categories which are often considered as helpful in achieving the changes needed to meet a particular overall objective.

Table 2.2. Policy instrument categories suitable and particularly relevant to stimulate specific changes.

	Changes								
Policy instruments	Total volume of activities	Technology: increasing economic efficiency	Technology: increasing environm. effect	Technology: shift to new process	Technology: shift to new products	(Re-) allocation: of inputs	(Re-) allocation: of outputs	Substitution on the level of inputs	Substitution on the level of outputs
direct regulation	*		*			*		*	
environmental legislation			*					*	
economic or financial instruments	*	*	*	*	*	*	*	*	
providing infrastructural facilities		*		*	*				
stimulating voluntary cooperation							*		*
moral suasion	*				*		*		*

Another important task is to link the combination of instruments and the desired change, i.e. the cells in Table 2.2, to specific methods. It is virtually impossible to do this unambiguously and to do it in such a way that consensus is attained. Table 2.3 presents a general indication of the variety of methods available; see Van den Bergh (1996).

From the point of view of establishing an intellectually sound and politically acceptable methodology for environmental policy analysis, there is the clear need to have explicit criteria for selecting evaluation methods (Janssen, 1992; Johansson, 1987; Munda *et al.*, 1994; Voogd, 1983). By comparing these criteria with the alternative approaches to environmental policy assessment, one obtains insight into the relative usefulness of alternate methods for different problems (Nijkamp and Blaas, 1994).

Table 2.3. A classification of important methods for environmental-economic research.

Statistical analysis	*Planning and optimization*	*Evaluation*
testing of hypotheses	programming models	cost-effectiveness
estimation of indicators	LP	analysis
valuation	integer	benefit-cost analysis
(controlled) experiments	nonlinear	multi-criteria analysis
econometric models	dynamic	multi-objective analysis
Partial analysis	discrete	*Decision theory*
marginal (price) analysis	discrete control theory	conflict analysis
single market analysis	continuous control	risk analysis
Systems analysis	theory	uncertainty analysis
macro models	*Spatial models*	expert analysis
input-output models	continuous spatial	rough set analysis
general equilibrium	interaction	scheduling
analysis	multiregional models	multi-criteria analysis
dynamic (stock-flow)	urban models	decision aid
models	network models	*Data analysis*
integrated models	zoning models	statistical analysis
cause-effect analysis	long run location	rough set analysis
hierarchical decision	analysis	artificial intelligence
models	land use models	
impact analysis		

The relevant selection criteria for evaluation methods can be subdivided into first- and second-order criteria. First-order criteria are mandatory criteria for the selection of an evaluation method; if a method does not comply with all first-order criteria which are relevant to a certain problem, this method cannot be applied to this problem. Second-order criteria refer to desirable features, but are not a priori mandatory criteria for the selection of an evaluation method. In practical terms, the problems of applying selection criteria relate to those desirable features and the implications for trade-offs which they entail. The aim is to find a method which complies with as many second-order criteria as possible.

One can also make a distinction between generic and specific criteria. Only part of the selection criteria is relevant to all evaluation issues. Most selection criteria are related to certain well defined environmental evaluation problems and, therefore, appraisal methods can only be judged in relation to the problem they are intended to solve. This means that the selection criteria related to certain well defined transport evaluation problems, for instance, have to be compared with the features of a number of available and relevant evaluation methods.

2.5 Spectrum of Policy Assessment Methods

Studies of relevance for policy support and preparation in the field of environmental economics have typically concentrated on the contents of single steps, and the use of methods in isolation. The actual implementation of these steps in a policy relevant context requires, however, also knowledge and experience about their interaction. This may involve feedback loops between steps, and affect the choice and reconsideration of the combination of tools in the sequence of steps. The outcomes of the steps depend on the type of issues or problems dealt with as well as the policy context in which they are being implemented.

To make the problem clearer, it can be stated as follows. The aim in environmental policy is to choose instruments, in a theoretical sense, to attain social optimality, and in a practical sense to ameliorate excessive environmental effects on health, living conditions, and the economy. The task comes down to providing a linkage between social objectives - economic, distributional, and environmental - and the mix of policy instruments. There are various combinations of tools available that can be used to create this link. What follows shows an economic perspective on linking methods, however, with a clear eye on the multidisciplinary character of the policy-strategic preparation context.

(A) An *impact analysis* for policy strategy is necessary as a first step prior to pursuing the other types of analyses. The aim is to determine the consequences of policy measures on a number of relevant variables. It is, however, not possible to distinguish between the time span of effects, and to optimize or aggregate without further methods such as under (B) to (F);

(B) A *partial equilibrium* policy-strategy analysis is used to estimate costs and benefits of specific policies taking only their immediate effects into account, and assuming away indirect effects (for example, as a result of relative price changes). These costs and benefits may be evaluated via benefit-cost evaluation.

(C) A *general equilibrium* policy-strategy analysis is based on an applied general equilibrium modelling approach that can deal with economic incentive systems, such as taxes, subsidies and tradeable permits, and can incorporate a considerable amount of the complexity present in market systems. It allows, therefore, for cost shifting effects being transmitted in a multisectoral market equilibrium setting.

(D) An *optimization-integrated* policy-strategy analysis is based on a constrained optimization model and integrates economic and environmental variables and constraints, to incorporate physical effects, processes and restrictions as well as economic-environmental interactions.

(E) A *scenario-integrated* policy-strategy analysis is based on a (dynamic) integrated economic-environmental model that generates values (over time) of performance indicator variables. These values can serve as an input of evaluation methods, notably focusing on multicriteria evaluation.

(F) A *scenario-optimization-integrated* policy-strategy analysis combines features of (D) and (E) in a heuristic, possibly iterative, procedure which may use an integrated system of descriptive models, scenario simulation, and optimization procedures. Multiple dimensions may imply the use of separate, sequential constrained optimizations, or be evaluated in a more integrated multicriteria method.

(A) proceeds any type of analysis; (B) and (C) are in the domain of traditional economic analysis, and are useful to decide about economic efficiency, least cost measures, and optimal tax systems; (D) and (E) can be considered as operations research type of methods valuable for dealing with planning, optimal (long run) growth, development and other dynamic issues, regional or even lower spatial levels of processes, and analyses on a material-physical level. The system of methods mentioned under (F) is seldomly employed in a systematic, planned and interactive manner. It has the attraction of combining functional features of (A) to (E). The Dutch environment-employment-multisectoral study for the Scientific Council for Government Policy (WRR) in 1987, and the Sporades Islands study by Giaoutzi and Nijkamp (1993) (see also Van den Bergh, 1996) is a good example of the latter case.

2.6 Assessment and Synthesis

The application of assessment procedures to environmental issues is complicated. Nevertheless, over the years many assessments have been conducted (Hedges and Olkin, 1985). These vary in their level of sophistication, in the details of their approach and, often, in their inherent objectivity. In other words, there is already a considerable and expanding body of knowledge upon which we may try to gain additional insights to assist in the development and improvement of appraisal methods. This is true even beyond the confines of ecological analysis (Oswald, 1991). Of course, gaps in our information frame remain and many aspects of the procedure need further original research but there is, nevertheless, a significant body of existing knowledge which may fruitfully be mined.

Meta-analysis essentially offers a range of techniques designed to generate additional information from an existing body of knowledge; it involves synthesis of empirical analysis to assess common features and variations across a range of prior studies (Hunter *et al.*, 1982; Fiske, 1983; Glass, 1976; Windle, 1994). Having looked into the nature of environmental evaluation problems, the question emerges as to the role that meta-policy analysis might play in helping improve our understanding of environmental policy measures.

While relatively little used in environmental research to-date, meta-analysis does have the potential to offer new insights into a number of important areas (Button, 1995). Its strength lies not in originality *per se*, but rather in extracting additional information from work which has already been done. It allows for the useful consideration of the pool of existing work and studies constructed on environmental issues and to draw from this pool common threads, outliers and linkages. From a more pragmatic perspective, it can help to isolate fields of study where further, original analysis could most cost-effectively be concentrated on. This advantage is a not an insignificant consideration in times of more restrictive research budgets.

A loose list, which is in no way meant to be exhaustive, of the types of environmental management and evaluation issues which might be addressed using meta-analysis can be defined within the boundaries of the following criteria.

There must be an existing body of studies which can be subjected to statistical procedures in the broadest meaning of the term. The number of studies need not be large (e.g. some medical meta-analyses have involved as few as three studies), but ideally it should not be too small. Normal statistical criteria favour the use of as many observations as possible.

There may be various objectives behind the use of meta-analysis in environmental policy assessment. This may be in terms, for instance, of attempting to seek out the common treats of previous successful packages of environmental policy measures or it may be in terms of looking for a summary measure from a body of prior analysis (e.g. relating to the damage done to property by particular atmospheric pollution). The need for this clarity of objective relates, in part, to ensuring the minimum bias introduced when selecting the studies for inclusion.

There must be a degree of commonality in the prior information to be examined. This may, for example, be spatial, temporal or subject-specific, but without this the realm of analysis would be excessively diverse. Defining the extent of commonality is itself a potential problem and is inevitably subjective. In some cases the problem of diversity may be contained, if it is possible to isolate the peculiarities of studies and to normalise them in the meta-analysis itself.

The subject matter of the study must be such that it can be handled within a formal, statistical or mathematical framework, although given the nature of work in the environmental field this may require the application of what might generally be termed 'soft modelling' (Nijkamp *et al.*, 1984). The modelling, however, must go beyond basic data description and elementary statistical tabulations, as there is a need for a more rigorous analysis of various distinct experiences.

2.7 Conclusions

There is a growing number of studies looking at the implications of alternative policy approaches to environmental-economic issues. Many fields in environmental economics are gradually building up a record of both empirical findings and of experiences in using alternative analytical techniques. The need to more fully embrace environmental considerations in policy-making is now widely accepted within the developed countries and is becoming increasingly institutionalised in transnational groupings such as the EU. The practical problems of doing this remain quite formidable, although significant progress has been made in recent years. This chapter has sought to offer a typology of the issues involved offering a policy framework that allows to examine how one can extract additional information from prior studies and policy analyses. After having provided a context under which efforts to bring environmental considerations into assessment and policy evaluation procedures could be well structured and integrated, it looked into the possibilities of applying meta-analysis in this field of research.

As a methodology, meta-analysis seems potentially suitable for this form of work. It is really the policy types of issue which require some additional thought on such matters as the use of soft modelling techniques. Several issues are of importance here. First, what are the behavioural responses to policies. This may be based on estimation of elasticities. Second, what are the environmental effects of policies. This may be based on time-series data or cross-country or cross-region comparisons. The choice of policies depends on its type, sector and country. Third, how can different investigations of hypothesis about historical patterns be meta-analyzed.

Meta-analytical methods may also be developed for dealing with alternative tasks, such as comparing, evaluating and ranking instruments, on the basis of well-defined criteria or goal functions, e.g., using multi-criteria evaluation methods. It will enable us to draw some additional insights from a collection of analyses in related policy fields which have already been completed. In the field of environmental analysis, and in particular that of environmental impact assessment, there is a considerable body of case study and similar material which is suitable for treatment using meta-analytical techniques. The new insights obtained from meta-analysis may be in the form of additional information of immediate use or it may highlight ways where further work would be helpful for decision making.

CHAPTER 3 META-APPROACHES: METHODOLOGICAL REMARKS

3.1 Introduction

Meta-analysis seems to be an appealing approach, but despite its potential there are several questions of a methodological nature which need to be addressed in greater detail. In this chapter the methodological issues of meta-approaches for environmental case studies are discussed. First, it deals with the intrinsic methodological complexity inherent to meta-approaches. This complexity is closely linked with the high level of transversality characterizing meta-analysis, both horizontally - especially identification, selection and analysis of the case studies under consideration - and vertically - notably identification and description of the problem to be studied, definition of the objectives of the study, and practical use of the results obtained. Six different levels of analysis can be distinguished in a meta-analytical approach; for each of these levels the characteristic methodological issues will be identified. In addition, the chapter sets out the range of alternative meta-analytical techniques. Finally, some concluding remarks are offered.

3.2 Methodological Complexity

To illustrate and tackle the fundamental issues connected with the specific methodological complexities inherent in meta-approaches, six different levels of analysis can be identified. These are logically connected and seem particularly relevant from a methodological point of view and for practical and interpretative reasons.

At first glance meta-approaches appear to offer considerable practical potential. This type of approach, however, is characterized by considerable methodological complexity, due not only to the specific objectives which are set in each case, but also to the intrinsic nature of the studies themselves. Transversality is important here, which refers to the intrinsic heterogeneous nature of the various studies and to the various different empirical or political processes or issues addressed in these studies. It is, therefore, not only necessary to make a suitable identification and selection of similar studies and to analyze them by means of an appropriate research technique (horizontal transversality), but also to adopt a vertical orientation, from the moment of identifying the specific problem to be studied to that of using the results of the study in a practical way or in a policy context.

In light of this, it is possible to identify different - somewhat stylized - analysis levels, each of which assumes a particular methodological importance:

(1) *Real-world level*: this indicates the space-time reality that is the context of all problems and phenomena studied, which through their interactions constitute a single, but complex system with many actors and issues involved.

(2) *Study* level: this consists of the identification, definition and description of the

problem selected. In general, this involves the formulation, explicit or implicit, of suitable theoretical hypotheses regarding the phenomenon studied; the verification of these hypotheses by means of the introduction of a pre-selected model; the use of suitable analysis techniques; the collection and/or elaboration of particular data; and the presentation - and usually also the interpretation and the analysis - of the results of this study.

(3) *Pre-meta-analysis* level: this consists of defining - explicitly and accurately - the object and the objectives of the meta-analysis to be carried out, and of indicating in particular the specific problems to be solved and the dimensions established in terms of time and space. The methods and techniques considered most suitable for carrying out successfully the planned research are also defined at this stage.

(4) *Study selection* level: once the object and objectives of the meta-analysis of a certain issue to be carried out have been established, it is necessary to determine and select - in terms of both quantity and quality - the individual studies to be reviewed, bearing in mind the meta-analytical techniques which are to be adopted and the ultimate aims of the synthesis itself.

(5) *Meta-analysis* level: at this level of comparative analysis the studies selected are analyzed thoroughly and critically by means of a proper formal technique; also the consistency of the results thus obtained is carefully evaluated, so as to offer a solution to the problem discussed, of which a proper synthesis is next presented.

(6) *Implementation* level: this constitutes a post-meta-analysis phase, a kind of feed-back or application to the real-world, which considers not only the expected results obtained by the synthetic study but also - and above all - the effects of the experience acquired; this phase supplies useful indications and practical suggestions (relevant to the problem studied but also to new studies to be carried out, to strategies to be adopted, etc.) which are not necessarily directly or closely connected with the original objectives of the analysis.

The list indicates that the steps prior to and during the application of meta-analysis have to be carefully judged and implemented. Clearly, not all these steps will necessarily have the same weight nor will be present in each study. It is conceivable that meta-analysis in general is characterized by a range of peculiar methodological complexities. These difficulties, moreover, become even more noticeable in the passage from application in the field of natural sciences to those in the field of social sciences in general, or environmental science in particular. To clarify this, the six different levels are discussed in turn.

3.3 The Real-world Level

With reference to the real-world level, it is noteworthy that, while many natural phenomena are characterized by a certain degree of regularity - more or less accentuated - and by a certain clear cause-effect relationship, the same is not always true for social phenomena. Here the cause-effect relationship is often marked by considerable uncertainty (see Chapter 6), since social phenomena do not depend solely on the nature of things but also on human behaviour (limited rationality, subjective preference, mutual influences, rigidities, etc.) and on the adoption of specific policies.

It is, therefore, easy to observe different characteristics of the same phenomenon or of the same problem, that is, non-random deviations of the data observed. These may depend on differences in times and places of observation; they may be a consequence of the presence of a number of conditioning factors, of different dimensions of the study carried out, or of the varying span of time required before the effects of certain social phenomena are felt. This is why, in the case of the social sciences, it is impossible to speak of strict experiments, and why it is necessary to use the terms quasi-experiments and quasi-scientific methods.

For these reasons, from a technical point of view, it is not always possible nor methodologically correct to carry out statistical analyses. The presence of a very limited number of stable laws, so-called statistical regularity, and difficulties involved in generation of samples which are random, independent and of equal dimensions, make it often hardly possible to make a correct and reliable use of descriptive statistics, of statistical, classical and Bayesian inference, or other techniques. It is often necessary to resort to modifications of the procedures employed and to the use of specific techniques which present serious limitations.

3.4 The Study Level

At the individual study level, the *a priori* problem is tackled by means of the application of the model, method and technique chosen, to the sample under study. With the social sciences therefore, it is possible that considerable interactions take place between the effective formulation of the problem faced and the data collected. If such data are incomplete, inadequate or insufficiently homogeneous, it may in fact be necessary to review the entire research project, including the definition itself of the problem faced. Moreover, the information available is often not of a quantitative nature and therefore not certain; this too leads to further and considerable methodological complications for the research project at hand due to the impossibility of making precise measurements.

The choice of the method to be adopted is clearly linked to the representation of the system and to the model selected. With reference to environmental systems analysis, there exists a fundamental type of uncertainty due to the limitation of observing and describing reality. In all sciences, theories or models are adopted, which may be interpreted as intellectually built artificial objects (Ravetz, 1971) in order to outline and represent selected aspects or dimensions of reality. Obtaining meaningful results depends on the craft skill with which these models are built. By means of a suitable application of these, one may then identify new properties and verify the extent to which they reflect reality.

This process may, therefore, be envisaged in circular form; starting from reality, where the phenomena exist, and continuing by induction to the world of the intellect, where theories and models are formulated, from which, by deduction, inferences may be drawn which, by means of an activity of validation, are then compared with real phenomena. It is therefore possible to observe (Miser, 1993) that, while the facts - as a result of careful and selected observations of reality - should prove completely objective, the induction, which transforms these facts into ideas, is a highly personal process, internal and value-loaded. The building of theories and models moreover, makes use of craft skills and imagination to combine the facts with other knowledge (complementary facts, well-established theories or models, etc.). The process of deduction makes use of logical formal instruments of a mathematical nature, in the widest sense of the word. The consequences, generated by deduction from the theories and models introduced, may then be checked against empirical phenomena from reality.

If we accept this concept of science in general, any theory or model is an approximation of a specific and selected portion of reality. This is true for all sciences, to the degree of the aspects of reality that they set out to approximate and summarize in their theory (Kemeny, 1959). This is consequently also true for environmental science and economics.

Furthermore, uncertainty will be found at the model level, where it is necessary to take decisions concerning the variables to be taken into account in the model itself: which of these should be taken into consideration, and which should be considered as indicators, or which are exogenous or endogenous? There are also other difficulties inherent in the formulation of suitable indicators, policies and instruments for the case study approach. Furthermore, there are problems of computation (techniques, nature of data, etc.), presentation, interpretation and discussion of the results of the individual study, which may naturally interact and create feed-back mechanisms.

3.5 The Pre-meta-analysis Level

In this phase the methodological choices faced are critical in terms of proportions, entity and difficulty, as they determine the progress and the quality of the synthesis to be carried out. This phase constitutes the first effective part of the synthesis, in which it is necessary to come to grips with a series of complex judgement problems of importance in order to guarantee a proper quality level for the synthesis aimed at.

Substantially, this phase consists of describing and defining the object and the peculiar objectives of the synthesis to be carried out, indicating the specific problems to be tackled and the dimensions established in terms of time and space. Subsequently, the methodologies and techniques considered most suitable for a correct completion of the program are then to be determined.

One implicit general objective in all syntheses, is that of reducing the level of subjectivity which is present in each individual study. This is particularly true in the field of the social sciences and environmental science, where there is a great variety of particular problems caused by the use of diverse methodologies in individual case studies; the varying degree of objectivity with which the information has been collected; the existence of studies which are not always well-designed and which do not always deal with appropriate questions; and a frequent difference in output measures and methodologies for summarizing data.

In this phase it is particularly useful to establish a correct methodological approach and, in particular, a suitable relationship between the principal objectives of the synthesis and the chosen techniques. The principal objectives of meta-analysis have already been mentioned. To reach these it is plausible to adopt a variety of techniques. It is clear that the choice of the meta-analytical techniques to be used is closely connected with the objectives of the synthesis.

If the principal object of the synthesis consists, for example, of reviewing and grouping a number of different studies characterized by a low level of homogeneity in order to identify the common traits, similar behaviour (spontaneous or induced) or complementarily of results, then traditional approaches of a qualitative type may be usefully employed. If the purpose of the synthesis is to identify relationships between the variables studied, evolutionary tendencies, synthetic indicators of calculated values, or other common elements which can be effectively described in quantitative terms, the use of suitable statistical techniques may be particularly appropriate. If the major aim of the analysis is a qualitative confrontation, a comparative evaluation or a ranking of individual studies, and if the results of each individual study may be referred to as criteria, then the use of quantitative or qualitative multi-criteria techniques may perhaps be suitable. If the single studies are to be grouped and classified according to some technical characteristics, data, results or, in general, in the presence of more than one attribute in order to obtain a clearer view on the analysis to be carried out, rough set analysis is likely to be the most suitable technique. If the aim is to identify one or more factors characterizing a particular study which cause the results to differ from those obtained from other rather similar studies, the use of techniques typical of the rough set approach may be particularly relevant. Finally, if the principal aim is the confrontation and evaluation of the particular methods and the different policies adopted with reference to the same problem in light of the results obtained from the successive individual case studies in order to acquire also some useful guidelines for treating in the future cases analogous to those studied, an approach based on multi-criteria techniques may be useful.

Clearly, also the nature of the results and their presentation will depend to a large extent on the technique employed.

3.6 The Study Selection Level

The choice of the individual studies to be analyzed is in general difficult. In a certain sense, this phase has the function of an "interface" between the preceding level and the one following, and may to a large extent condition the results of the synthetic meta-analysis study. Although this problem exists for any type of meta-analysis, it is more common in studies of a socio-economic type, where it is only possible to speak of "quasi-experiments", due to the absence of stable laws governing social phenomena, the use of research methodologies which vary greatly, and the absence of standardization.

From an operational point of view, it is necessary to decide how many and which single studies should be taken into consideration. Methodologically and ideally, this would require:

- defining the group of all individual studies of the given problem which are more or less similar and may be considered eligible;
- establishing selection criteria;
- deciding on the number of single studies to be analyzed;
- making a selection.

These phases cannot be considered as completely separate from one another. For example, if the number of studies considered eligible is reduced, this will have a drastic effect on all subsequent phases. With regard to those studies which are potential candidates, a first problem which can be prejudicial stems from the fact that the individual studies considered eligible are for the most part those already published; furthermore, especially in certain scientific areas, these are exclusively studies in which the results obtained have been positive or confirmative regarding some underlying assumptions. It therefore follows that it is very difficult to take into consideration those scientific studies which may be absolutely correct from a methodological point of view, but which have not reached positive results or a clear confirmation (Wachter, 1988; *The Economist*, 1991). This tendency towards a continuity with consolidated results may render meta-analytical research in many fields less objective, since it excludes a priori the possibility of considering results which do not conform to the prevailing line of thought.

The same problem may also arise for those individual studies of great interest which, irrespective of the quality of the results, are not published, or are published in such a way that they remain inaccessible for meta-analytical review. This is particularly true in the field of applied economics, where, for example, many studies are carried out as private consultancy or in any case in restricted and confidential circles (the "grey circuit"); at times only some of the principal results of the study are available, without the fundamental methodological details.

With regard to the criteria for the selection of the individual studies, particular care must be taken to ensure their similarity; ideally, studies should be chosen which differ in only one fundamental characteristic (by way of controlled experiment), but as this is practically impossible, it is necessary to consider those studies which differ in as few characteristics as possible. An idea of the principal factors characterizing the individual studies is offered in Van den Bergh *et al.* (1995a).

Besides these intrinsic characteristics, an important comparison regards the different sequences or steps in which the studies were performed. This comparison aims to indicate whether the studies considered are sufficiently similar from a methodological point of view, so that their results may be considered more or less comparable. It is, strictly speaking, impossible to carry out any form of comparative evaluation, classification or synthesis of the results, unless these are held to be sufficiently uniform from a methodological point of view and sufficiently standardized in their presentation. The principal aim of the similarity analysis of studies is to make a preliminary selection or a preliminary classification. But the real selection of the studies takes place subsequently, taking into account the specific problem to be considered, their precise purposes, and the methods and techniques to be used in the synthesis concerned. These are the main criteria to be borne in mind in the final choice of the studies to be analyzed. For instance, the choice whether to employ statistical techniques or quantitative analyses can have a drastic influence on the choice of individual studies.

The problem of selection is also closely linked to that of the number of studies to be selected. It is possible to decide on a plausible choice among the existing studies but it may also be useful to analyze the complete range of individual studies available, or some random samples of these. This choice depends above all on the technique decided on for the study of the synthesis at hand, on the total number of individual studies available and on the intrinsic characteristics of these.

Clearly, it is sometimes relevant to undertake some classification of all existing studies on the basis of one or more of the intrinsic characteristics chosen or on the basis of other factors which have been defined and evaluated. These classifications may prove to be very useful in the difficult selection phase.

Finally, with reference to the need for uniformity and standardization, or in order to improve the estimation efficiency in the use of quantitative techniques, it may be necessary to perform further experiments or simulations (Koslowsky and Sagie, 1993) or to carry out new elaborations, calculations or estimations of the data presented in the individual studies. This may arise because of technical errors in computing, or from an evaluation of the role of statistical artifacts, sampling errors and unreliability of measurements (Ones *et al.*, 1994; Witt and Nye, 1992); alternatively it may stem from a comparability of research findings utilizing different data collection forms, or in general, from the need to reduce as far as possible the subjectivity inherent in the individual studies, especially in the social sciences or from the need to increase the quantity of available data. For this purpose it can be useful to build taxonomies for organizing data, using, in some cases, multidimensional frameworks.

Moreover, in this phase, especially when some quantitative techniques of meta-analysis are used, a lack of a sufficient homogeneity or incompleteness of the data available may occur. In other words, an analysis of the individual studies selected may immediately reveal the need to manipulate in some way the data supplied by these studies in order to render them comparable or suitable for elaboration with the chosen methodology. These manipulations, which should in any case be "neutral", must be performed only if they prove absolutely necessary in order to guarantee the methodological precision and scientific accuracy of the results to be achieved in the synthesis, and should in any case be adequately and explicitly shown.

Furthermore, the problem may arise that some desired data, important for a more detailed and accurate study of the problem to be tackled or absolutely essential for the use of certain quantitative techniques which require this data to be complete, proves to be partially or totally lacking in some of the individual studies. In this case, bearing in mind the importance of the information contained in this data and the number of individual studies available, one must choose between considering only the original data included in the various studies (original design); reducing the data of all the studies, taking into consideration the data remaining after the intersection of these (reduced design); completing the missing data by calculation, or by estimating it in light of the information available (full design).

The choice between options is sometimes enforced by a technical need for completeness, or influenced by the advisability of not losing information. Certainly, when available data are a result of some hypotheses which are not substantiated by the more or less explicit information contained in the individual study considered, it is preferable to follow the original design or, to use the reduced design. If, on the other hand, the importance of not losing crucial information is evident, in connection with the suppression of partially incomplete data in some of the individual studies, one could resort to the full design, while supporting the incomplete data with suitable elaborations which, in the absence of precise information on the subject, must ensure the neutrality of these. In other words, the completion of data permits only the use of the chosen techniques, without imposing however options which are unjustified or not explicitly recognized by the researcher. It is then preferable to obtain synthetic results which are strong and methodologically correct, but perhaps less complete, rather than to supply more detailed and suggestive evidence based on hypotheses of a debatable validity.

3.7 The Meta-analysis Level

The central aim of the studies carried out at a meta-analysis level is to conduct a synthesis of the studies selected at the previous level. In this phase, the main methodological problem is the choice of the meta-analytical technique to be adopted.

In general, in this important stage, it is necessary to make an accurate analysis of each of the individual studies selected, recognizing all alterations or aggregations made, considering the most significant methodological aspects of these studies, identifying the basic results, and pulling together all results obtained from the analysis of each individual study. These findings must then be compared to make an efficacious synthesis with reference to the problem faced, to evaluate the consistency and to analyze the robustness of survey conditions, and, if needed, to carry out suitable feed-back analyses with respect to the previous levels.

To produce this synthesis, it is possible to use a variety of techniques. The most important of these are: traditional review, content analysis, statistically-based meta-analysis, meta-multicriteria analysis, epistemological analysis, uncertainty analysis (Funtowicz and Ravetz, 1987, 1990) and rough set analysis. Such approaches are discussed in more detail in Section 3.9. The traditional approaches of a literary type suffer from several limitations. Outputs in the form of taxonomies, for example, without any specific attempt to relate these to the purpose of the review, constitute merely a description of the problem. Traditional reviews also are generally highly subjective. There is often no evidence of the degree of conflict of the results analyzed; selection of individual studies with similar conclusions is often done without considering the quality of the data or of the techniques used and serious difficulties can arise in mentally handling a large number of different findings.

The use of quantitative techniques, and statistics in particular, can at least in part reduce these problems, ensuring a greater consistency in the results obtained. To use these techniques however, it is necessary for the results of the individual studies to be complete and have a high degree of homogeneity. There are characteristics which are not always found, especially in socio-economic research, except at the expense of making more or less arbitrary adjustments. In the choice of the technique to be used, it is, therefore, necessary to bear in mind the principle that it is not possible to generate information not contained in the results of the individual studies, while, at the same time, one should make use of all information that these results are able to offer.

In conclusion, the selection and implementation of the most suitable meta-analytical techniques cannot be treated in isolation from the context of the whole analysis, but must be made by taking into account all aspects of the problem faced. Nor is the simultaneous use of different approaches to be dismissed *a priori*. At times, it may prove extremely useful to carry out the study by using a variety of meta-analytical techniques, even of a completely different nature (Woodside *et al.*, 1993). The failure or success of the use of one of such techniques, and in particular the understanding of the reasons for this, can prove extremely useful. Sometimes divergent conclusions, for example, stem from variations in goals rather than from variations in the studies selected to be reviewed.

3.8 The Implementation Level

Regarding meta-analysis in environmental management, it is useful to distinguish three different classes of problems on the basis of their goals (Miser, 1993):

(1) For scientific problems the goal is to solve the problem. This means arriving at intellectual objects that are adequate approximations of reality. While the function to be performed by the solution is to contribute new results to the field; this may involve comparing and ranking studies, correlating aggregated data with other characteristics of each study, identifying moderator variables, etc.

(2) For technical problems, the function to be performed sets the problem. The task is accomplished, if the solution enables the function to be performed; this may involve summarizing indicators, averaging estimated values and parameters, etc.

(3) For practical problems the goal is to serve policy purposes, and the problem is solved when a means for serving these purposes has been devised and shown to be effective. This may involve evaluating and ranking different methods or alternative policies, aggregating studies by considering complementary results or perspectives, etc.

At the implementation level various study effects are observed which may go beyond the basic results expected from it. This phase may assume great importance, especially in some scientific or practical problems where the implementation of a technique, the adoption of a strategy or the choice of a practical approach is desirable.

More generally, it can be said that the result to be strived for from any meta-analytical study is a "learning by comparing". In other words, the acquisition of a new experience which is immediately applicable elsewhere in the scientific field. This could comprise, for example, the design of a new theory, elaboration of the existing explanations, and identification of areas with specific studies or fruitful additional research. In the applications, for instance, it may entail developing effective strategies or guidelines for effective management and indicating types of behaviour consistent with certain objectives.

3.9 Alternative Meta-analytical Techniques

There are a variety of techniques which can be employed in a meta-analysis; the adoption of one will be influenced by the particular types of insight being sought. These include:

(1) Summarising over a collection of similar studies relationships, indicators and so on.
(2) Averaging, possibly using weights, for collections of values obtained in similar studies.
(3) Comparing, evaluating and ranking studies on the basis of well-defined criteria or goal functions.
(4) Aggregating studies, by taking complementary results or perspectives.
(5) Apprehending common elements in different studies.
(6) Comparing outcomes of different methods applied to similar questions.
(7) Tracing factors that are responsible for differing results across similar studies.

We will concisely describe here four classes of meta-analysis techniques. First, conventional statistical methods may be used which have the advantages that they are well understood by economists and have been extensively used in meta-analysis work in the natural sciences. These techniques can play useful roles when concern is with purposes 1,2,4,5, 6 and 7 (Cook and Leviton, 1980; Hunter *et al.*, 1982; Hedges and Olkin, 1985; Light and Pillemer, 1984; Mann, 1990; Rosenthal, 1991; Button, 1994 and 1995, Wachter, 1988; and Wolf, 1986). The problem is that they require information to be presented in a manner which is amenable to statistical analysis and this may limit both the range of topics which can be treated or the types of effect that can be considered across a range of studies. Discrete modelling procedures have, in recent years, reduced this limitation but it still remains.

Second, there are various forms of meta-multicriteria analysis (MCA). These can be very useful for steps 3 and 6, especially where multiple criteria, or objectives, may underlie a comparison of different, but similar studies. This is especially relevant when studies do not offer a simple estimation of an important indicator or where the interest lies in looking at a number of indicators (Keeney and Raiffa, 1976; Rietveld, 1980; Chankong and Haimes, 1983; Yu, 1985; Steuer, 1986; Zeleny, 1982; and Nijkamp *et al.*, 1990). For applications of multicriteria techniques in the field of environmental management and policy, see Janssen (1992), and for uncertainty and multicriteria analysis in the context of environmental issues Munda (1993).

Third, there is epistemological or expert analysis, such as the Delphi method or Numeral-Unit-Spread-Assessment-Pedigree (NUSAP) approach. In the context of ecological economics and policy, NUSAP has been suggested by Funtowicz and Ravetz (1990) as a potentially useful approach. The NUSAP scheme was designed as a robust system of notations for expressing and communicating uncertainty in quantitative information. The basic idea is that when either system's uncertainties or decision stakes are large, traditional applied science is inadequate, and a new approach is required in judging, evaluating or summarizing existing insights. Funtowicz and Ravetz refer to an "extended peer community" which is to involve all stake-holders related to the respective situation or issue, including citizens, experts and policy-makers. It should allow for a multi-dimensional, multi-perspective approach, which explicitly treats emergent complexity of systems. It allows meta-analysis to deal with policy studies which are not limited to such things as supply of unidimensional estimates, or optimal patterns, but cover a whole range of issues of a qualitative and complex nature, including trade-off and distributional issues.

Finally, there is rough set analysis which offers the potential for handling a much more diverse and less immediately tangible set of factors. Rough set analysis, proposed in the early 1980's by Pawlak (1982), provides a formal tool for transforming a data set, such as comprising past examples or a record of experience, into knowledge, in the sense of ability to classify objects. In resulting classes it is not possible to distinguish objects on the basis of given information about them. This imperfect information causes indiscernibility of objects through the values of the attributes describing them and prevents their precise assignment to a set. In this case the only sets which can be characterized in terms of such attributes are lower and upper approximations of the set of objects. Rough set theory has proved to be a useful tool with respect to a large class of multi-attribute decision problems. It can deal in an effective way with problems of explanation and prescription of a decision situation where knowledge is imperfect. It can help to evaluate the importance of particular attributes and elimination of redundant ones from a decision table, and generate sorting rules using only the remaining attributes so as to support new decisions in policy.

Whatever technique used, the quality of its outcomes will depend on measurement characteristics on two levels: of the original studies which are being meta-evaluated; and of the translation of uncertainties inherent in interpreting study outcomes on the meta-analytical level. Sometimes additional information may be present about uncertainty on the level of specific studies involved in the meta-analytical process. However, if this is not the case, one should adopt an *ad-hoc* approach to cover this uncertainty somehow.

3.10 Concluding Remarks

In describing the process outlined in Sections 3.2 to 3.8, one may consider the claim by Yu (1990) that through decision analysis an enlargement of the "competence set" is obtained, or an expansion of the "habitual domain"; these lead each individual to a better comprehension of reality and to a more fruitful approach to all decision problems, especially those which are new and complex, by forming so-called winning strategies. This appears to be of particular importance in the field of environmental science, where the policies to be adopted and the decisions to be made are particularly difficult, because of the unpredictability and uncontrollability of processes in the interaction between man and his environment. As a result, it is sometimes hard to speak of structured decision problems and to require not only the use of suitable highly-sophisticated techniques of decision aid, but also of approaches which offer the possibility to understand better the existing relations between natural phenomena, human behaviour, policy instruments and economic policies. To avoid the trap of methodological complexity and low level information, the meta-level approach is a promising new direction in environmental policy and impact assessment.

CHAPTER 4 REVIEW OF META-ANALYSIS WITH APPLICATIONS TO ECONOMICS

4.1 Origin of Meta-analysis

After the previous introductory chapters we will now treat in more detail the essence of meta-analysis. The term meta-analysis was first introduced by Glass in 1976, although the idea of analyzing statistically large quantities of scientific publications dates back from the early 1970s. Glass (1976) provides a widely accepted definition: "Meta-analysis refers to the statistical analysis of a large collection of results from individual studies for the purpose of integrating the findings. It connotes a rigorous alternative to the casual, narrative discussions of research studies which typify our attempts to make sense of the rapidly expanding research literature."

Other definitions are provided by several sources. For example, an article in *The Economist* (1991) states: "Meta-analysis compares the deviations found in lots of studies of the same thing. If the deviations are randomly scattered around zero then there is no effect. But if they cluster off to one side, the meta-analysis shows that something is going on, even though the individual results may not be significant in themselves." Cook (1992) describes meta-analysis as follows: "Meta-analysis offers a set of quantitative techniques that permit synthesizing results of many types of research, including opinion surveys, correlational studies, experimental and quasi-experimental studies and regression analysis probing causal models." In an earlier publication, Glass *et al.* (1984) had already claimed: "Meta-analysis is data-analysis applied to quantitative summaries of individual experiments. It is the statistical analysis of the summary findings of many empirical studies." And finally, Pettiti (1994) describes meta-analysis as follows: "The overall goal of meta-analysis is to combine the results of previous studies to arrive at summary conclusions about a body of research. It is most useful in summarizing prior research when individual studies are small and they are individually too small to yield a valid conclusion."

In general, the meta-analytical approaches provide a series of techniques that allow the cumulative results of a set of individual studies to be pulled together. It permits a quantitative aggregation of results across studies. In doing this, it can not only help to provide more accurate evaluations of quantitative parameters but may also offer insights into phenomena for which no specific study currently exists. It can also, in certain circumstances, help to pinpoint political bias and provide more clearly defined valuations of the economic costs and benefits from the plethora of data that exists. It can act as a supplement to more common literary type approaches when reviewing the usefulness of parameters derived from prior studies and help direct new research to areas where there is greatest need. In this context, Hunter *et al.* (1982) argue that "What is needed are methods that will integrate results from existing studies to reveal patterns of relatively invariant underlying relations and causalities, the establishment of which will constitute general principles and cumulative knowledge."

Recently there has been a gradual increase of empirical research in economics and environmental research, using microeconomic data for the assessment of policy proposals and plans, in both the private and the public sector (Oswald, 1991). In part this has been achieved through the increased use of basic quantitative analysis although recently advances in statistical procedures have also seen rigorous techniques of a more qualitative nature being adopted (e.g. multi-criteria analysis). Advances in computational power have been one of the driving forces behind this as the costs of estimating economic parameters have fallen and large data sets are increasingly becoming available. Whether the new insights we have gained have been cost-effective, however, in terms of the additions they make to human knowledge is a question we avoid.

The need to make better use of existing research is not only an issue in economics, but applies to all disciplines. It may be illuminating to refer here to Kissinger (1960) who argued that: "The production of so much research often simply adds another burden to already overworked officials [...]. Few if any of the recent crises of US policy have been caused by the unavailability of data. Our policy makers do not lack in advice; they are in many respects overwhelmed by it." Meta-analysis, while certainly not providing a basis to meet all current problems encountered in forecasting economic impacts, would seem to offer at least a partial way forward.

The aim of this chapter is to explore ways in which greater insight may be gained by remining and reassessing the findings of previous studies in the field of economics. In doing this it draws upon a range of statistical synthesis studies for illustration and looks at some of the findings obtained as well as assessing the methods used.

Of course reviewing and discussing earlier work in a literary manner has always held a central place in the social sciences, and in economics the *Journal of Economic Literature* has long been at the forefront of this trend with the *Journal of Economic Surveys* now providing a valuable addition. Indeed, the *Journal of Economic Surveys* has carried a short paper on meta-analysis (Stanley and Jarrell, 1989) in the past. The tradition, however, has very much been a literary one. The surveys of empirical findings from earlier works, even those of quite sophisticated econometric analysis have tended to report findings in tabular or 'pictorial' fashion with verbal comment and discussion of the strengths and weaknesses of each study included. This approach is obviously valuable and important but with respect to empirical works such qualitative assessments inevitably often omit the full inclusion of important quantitative consideration. A complementary approach is to supplement the traditional literary methodology with that of the statistician. This is in essence what the subject of this chapter, meta-analysis, does; it provides a quantitative summary or assessment of research domains.

The emphasis of this, admittedly essentially literary, review is on looking at the usefulness of meta-analysis to improve our understanding of microeconomic issues and to consider some of the actual applications of the approach that have been attempted over recent years in the field. This is not to say that it has no potential role in macroeconomics for synthesising work in this field, especially since the growth in empirical work has been equally rapid. It has, for instance, been used to analyse the growing quantitative literature on links between infrastructure investment and economics growth (Button and Rietveld, 1996). Rather, the bias reflects the need to contain the assessment within tractable boundaries one of which is the authors' own comparative advantages in terms of their own research interests and backgrounds. In most ways, though, the general points made regarding the application of meta-analysis by microeconomists also hold for its use by their more aggregate inclined brethren.

4.2 The Perspective of Meta-analysis

Good surveys are useful. Any study of citations soon shows this. Surveys provide an immediate access point to a literature for those new to an area, they offer a useful source of up-dating material for those already involved and they serve as important reference points. Traditional surveys in economics have tended to be of a literary nature and these unquestionably have immense value. This chapter is concerned with looking at whether there are alternative ways in which information can be surveyed to complement the conventional approach. More specifically, it looks at the question of synthesising research statistically.

Meta-analysis offers a number of possible advantages or, perhaps more accurately, different perspectives when compared to conventional procedures which follow a narrative style for bringing together information from previous case studies. In particular, the standard approach to reviewing previous work, and then making use of the selected results, suffers from several limitations some of which an appropriate meta-analysis can circumvent, minimise or at least, make transparant (Light and Pillemer, 1984). Although it is something of a straw man constructed to help exposition, we can compare meta-analysis with conventional reviewing techniques, by which we mean studies which list results of previous work and debate their pros and cons in a literary fashion with the aim of isolating superior work, or analysis which may be used as the basis for further analysis. In a sense conventional surveys can be seen as seeking out broad guidelines and trends. They contain extensive listings, tables, taxonomies and so on but little, if any, detailed statistical synthesis of the results provided by the studies being reviewed.

Traditional surveys have also other limitations. First, there is the problem that the output of most traditional reviews tends to be in the form of taxonomies of findings without any specific attempt to relate these to the review's purpose. For instance, they may provide lists of estimated price elasticities according to whether they reflect long or short term responsiveness but offer no guidance as to when any particular elasticity is relevant. In consequence, such reviews seldom meet the needs of those engaged in activities which require quantitative forecasting. Added to this, is the often hidden subjectivity that tends to accompany a basically literary type approach. Further, the result of any statistical verification procedure can be a greater degree of conflict in the outcomes than exists in the base studies themselves. Although disagreements amongst findings is itself not bad - it suggests a need for seeking an explanation for the diversity - traditional methods do not normally attempt more than a description of the problem.

Second, there is the problem that traditional reviewing, because it does not necessarily embrace sound statistical practice, is usually scientifically unsound itself.

A common problem is that if a majority of studies come up with similar conclusions these are accepted on a sort of voting basis irrespective of the quality of the data used or reliability of techniques employed. Finally, the traditional review process is frequently inefficient because of the difficulties, both for authors and readers, of mentally handling a large number of different findings.

The idea of meta-analysis has a relatively extensive history, especially with respect to replicated experiments, and it has been widely employed in psychology and medical research. Its application to economic issues, though, has until relatively recently been extremely limited. Some of the few recent examples include the examination of causes of X-inefficiencies across a range of different industrial studies (Button and Weyman-Jones, 1994), of education (Phillips, 1994) and of absenteeism (Farrell and Stamm, 1988; Mitra *et al.*, 1992). The origins of the approach are generally traced back to agricultural experiments conducted during the early part of the twentieth century which initially saw statistical procedures being developed for independent experiments, but which then led on to methodologies being devised for combining the results of these experiments (e.g. Lush, 1931). The procedures were then gradually taken up in the medical field, and in particular in psychology.

The early work in the social sciences tended to be concentrated in social psychology and in areas such as research methodology and experimenter expectations (Rosenthal, 1963). The major academic impact on the social sciences can, perhaps, be traced to the work of Glass (1976), who also coined the term meta-analysis, on the effectiveness of psychotherapy. It was also taken up, to some extent, in the early 1970s by political scientists in the USA when trying to assess the implications of the New Frontier and Great Society programmes. The emphasis on social psychology has continued in much of the recent meta-analysis but there has gradually been an increase in work in the management science area - e.g. in the personnel field (Lowry, 1994; Robertson and Kinder, 1993); labour efficiency (Speilberger and Reheiser, 1994; Macan, 1994); marketing (Rust and Cooil, 1994; Sujan *et al.*, 1994; Szymanski *et al.*, 1993); leadership (Wofford and Liska, 1993) and in education (Orth and Martin, 1994; Wang *et al.*, 1993). The impact of meta-analysis has, though, been felt less in economics than in many other fields and this is clearly reflected in the illustrative material contained in the social science oriented texts on meta-analysis.

4.3 Meta-analysis in Context

While meta-analysis can provide a framework for tackling the difficulties of conventional procedures of reviewing previous work, it should be treated as an approach, rather than a single methodology. Indeed, the techniques involved in meta-analysis are diverse because of the wide ranging nature of the topics that can be addressed and the scope of the statistical issues that are thrown up when attempting an analysis of heterogeneous micro econometric works. A broad impression, however, of what is being attempted in the economics field is possible. In particular, the work that has deployed meta-analysis has frequently, but not exclusively, been concerned with moderator variables. Most of the work that has been conducted in this field has employed meta-regression analysis of the general form (Stanley and Jarrell, 1989):

$$b_j = \beta + \Sigma\alpha_k Z_{jk} + u_j \quad (j = 1,2, \ldots L) \quad (k = 1,2, \ldots M) \tag{4.1}$$

where b_j is the reported estimate of the relationship of interest in the *j*th study from a total of L studies; β is the summary value of b, Z_{jk} are variables that reflect the relevant characteristics of an empirical study that could explain variations amongst studies; α_k are the coefficients of the M different study characteristics which are controlled for and u_j is the error term.

Since the exact nature of each meta-analysis seeking moderator variables tends to be different, it is perhaps helpful in providing a little flavour of the types of variables which are often explored by taking a specific issue. For ease of exposition, therefore, we take a subject which is reviewed in more detail later, namely evaluating traffic noise nuisance. The general functional form which would seem appropriate when seeking meta-analysis moderator variables in this context can be summarised along the following lines:

$$Y = f\,(P, X, R, T, L) + \text{Error} \tag{4.2}$$

where:

- Y represents an outcome of interested. This may be a single measure, such dbA in the case of noise, or it may reflect a variety of differing effects such as level, duration and pitch,
- P can be treated as the specific cause of the problem (such as traffic levels and proximity to source),
- X represents features of those affected by the nuisance (such as their age and income),
- R, since the analysis is not based upon primary data, but rather the combination of other studies, represents the characteristics of the research methods used in each study (e.g. econometric or survey) and the data used (e.g. time series or cross-sectional),
- T indicates the period covered by each study to allow for any underlying dynamic effects (such as systematic changes in social preferences),
- L indicates the location of the study (which could be spatial, such as urban and rural, or may relate to the country forming the basis for each data set).

The right hand side of the equation effectively extends the argument of Hunter *et al.* (1982) that much of the observed variance in correlations across studies can be accounted for by three statistical artefacts: unreliable data due to small sample size, inter-study differences in the reliability with which dependent and independent variables are measured, and inter-study differences in restrictions of range.

A few words of justification for this type of framework are necessary. First, much of the work using meta-analysis in the physical sciences tends to focus on P and X in the above listing. This is mainly because of the ability of those working in these fields to compare strict experiments where methodologies are identical and, also results tend to be reported in a rather more standardised way. The importance of, so-called, artefact effects, which are normally embraced in R, in the social sciences, where the need for quasi-experimentation has often resulted in a considerable range of diverse model specifications, information gathering procedures and econometric estimation methods being adopted, would seem to be much greater. Winston (1985) provides some illustrations of this. The same argument applies to the variables T and L. This obviously poses practical problems when attempting to express the various studies' approaches which is consistent with the inherently numerical variables which are the basis of most scientific meta-analysis.

Much of the early work using meta techniques has been concerned with quantifiable, or at least quasi-quantifiable, effects (Hedges and Olkin, 1985; Hunter *et al.*, 1982; Rosenthal, 1991). For example, in terms of the effectiveness of various medical treatments in prolonging life or improving the performance of bodily or mental functioning. Further, although meta-analysis has tended to be used in cause-effect types of experimental situations, it has also been adapted to circumstances where there has been felt a need to summarise particular phenomena. In some uses this has involved work in the social sciences such as economics, for instance, the estimation of point values (e.g. the demand elasticity of a rise in gasoline tax) and of the environmental and land-use implications of transport.

This early work can further be typified by its attention on three broad issues or purposes (Rosenthal, 1991):

- Summarising for a set of studies what the overall relationship is between two variables investigated in each study. The term 'variables' is being used in the most general sense. It may be thought of in traditional terms, say price and quantity, but equally it could be considered as policy changes, such as deregulation, and social impacts. These essentially produce the average correlation found in all the studies summarised. In the case of economics, for instance, a modern example might be estimating the appropriate average price elasticity of demand for some commodity from a series of small econometric studies.
- Looking at the factors which are associated with variations in the nature of relationships between two variables over a range of studies (referred to as moderator variables). Guzzo *et al.* (1987) look at this aspect of meta-analysis in more detail. This yields a correlation between some characteristic of the studies and the correlation found within the studies. In the elasticities example this might involve looking at why elasticities differ as between individual studies, say due to income variations or the time frame examined.
- Looking at the aggregate data for each study and correlating this with other characteristics of the study. The aim is often to look at these average figures to gain insights into ways subsequent empirical work might be conducted. Remaining with the elasticities example, this may involve looking at changes in income over time and seeing whether cyclical variables would enhance future econometric work.

These three different purposes can be illustrated by taking the very simple example on the performance of business firms set out in Table 4.1. This hypothetical table provides the results of five studies looking at the internal forecasts and actual performance profit of different samples of companies. If interest is in the first purpose, the summarising function, of meta-analysis, then the focus would be on column A and on estimating some summary statistic (e.g. the mean correlation figure) across the studies. The determination of moderator variables, if the second purpose is the underlying rationale, would mean looking at the correlation between the data in columns A and B (r = 0.911) and/or A and C (r = 0.969). In this hypothetical case the suggestion is that higher profits, rather than the employment of economists, are associated with more accurate profit forecasts. Finally, if the purpose is of the third type, then the aggregate analysis would look at the correlation between columns B and C (r = 0.890) which suggests links between profit rates and the employment of economists (although the direction of causation is not clear). In this context, further work may, therefore, require more thinking about whether economists improve the forecasts of profits or the actual profit levels of firms.

Table 4.1. Hypothetical example of different uses of meta-analysis.

Study	A Correlation between forecast and actual profits	B Mean number of trained economists employed	C Mean rate of profits
1	0.12	5	3%
2	0.72	12	17%
3	0.99	18	19%
4	0.34	4	9%
5	0.45	4	8%
Mean	0.52		

While there has been a gradual development in the meta-analytical approach, the three purposes outlined above still form the underlying basis of what is being done; with studies in the microeconomics field focusing mainly on the first two. The methods used are gradually being refined. What has happened is, for example, that improved statistical techniques have permitted a wider set of variables to be considered beyond the bivariate approaches of the pioneering work, and have enhanced the sophistication with which the analysis is now conducted.

The question of estimation methodology may be also thought to be particularly important in many sub-areas of economics which are now expanding. This is the case, for example, with several aspects of the environmental economics because different valuation procedures can embrace differing components of cost (Johansson, 1987; Pearce and Markandya, 1989; Freeman, 1993). In part this is due to the elements that make up the external cost of many environmental considerations. Boulding and Lundstedt (1988), for example, argue that individuals values influence their behaviour in at least three ways, through observed choice (i.e. revealed preference), peoples' conversations (i.e. stated preference) and adaptations as part of the learning process. Smith (1989) suggests that all three provides insights into measuring the economic value placed on environmental amenities although there is no reason to expect the values to be the same. A simpler breakdown argues that the individuals' preference for environmental protection consists of three components:

Total Economic Value = User Value + Option Value + Existence Value (4.3)

User value here being that which may be observed from behaviour; option value is the result of individuals desiring to keep an environmental asset in case they wish to use it in the future, while existence value is the importance individuals attach to the very existence of an asset even if they never have or will have personal contact with it. These categories can be further refined but from our perspective the point is that estimation techniques that rely only on revealed preference techniques tend to understate option and existence values. Stated preference techniques can embrace them more fully and thus generally yield a higher valuation, apart from other biases. Similar types of issues emerge in other areas of economics.

The same sorts of argument apply to T and L, especially when time series and cross-sectional work is combined in the meta-analysis or there is considerable geographical variation in the sources of the data used. The importance of the time the studies were conducted and their and locations stem from the underlying economic preference model. At one level preferences are context specific and differing contexts may result in different valuations. One element is clearly information. Again, on the environmental theme, once lead additive to fuel was felt to have negligible environmental effects but medical work in the 1960s brought this into question and, thus, preferences changed. We are now moving back in the other direction. Equally, local conditions can be important because they can influence the choice set. A wealthy country provides a richer choice set than a poorer one and preferences may be affected. Added to this must be the constraints imposed by political and technical environments which are often space specific.

4.4 Problems of Employing Meta-analysis in Economic Research

4.4.1 General problems

The general problems of using meta-analysis can be viewed, first, in terms of the objectivity with which the body of information is collected and, secondly, in terms of the assumption that there exists a set of well-designed studies that address the appropriate questions, use similar output measures and focus on the methodology needed for summarising the data.

While meta-analysis can reduce subjectivity in, for instance, assessing different evaluation studies and seeking out relevant common threads, it cannot remove it. This is so because the technique brings together a number of studies and the analyst is obviously still instrumental in their selection. In this sense, subjectivity is inherent but, equally, the use of statistical analysis does mean that the conclusions drawn from any given set of studies can be subjected to more rigorous analysis.

Perhaps a more serious practical problem is the possible bias that results from the nature of the studies which are included or excluded. Looking across a range of social science meta-analyses a variety of reasons have been given for excluding some studies from the analysis. Nelson (1980) looks only at what he considers the best studies, Button and Weyman-Jones (1994) limit themselves to studies which pass an arbitrary statistical cut-off point and Mitra *et al.* (1992) excluded non-published studies and those which could not be converted into a usable form while one was set aside because of space limitations. Smith and Kaoru (1990a) were forced to restrict themselves to only a relatively small number of the studies they had found mainly because of incompatibility in reporting problems.

Linked to this in the particular case of the microeconomic literature, is the prevailing tendency only to achieve publication for academic work containing positive results. Mode choice studies in transport economics with an insignificant value for travel time costs, for example, seldom make an appearance in refereed journals. This makes it difficult to incorporate quite legitimate negative results in any overview analysis.

This inclusion issue is also linked to the broader assumptions of meta-analysis concerning the separability of studies, and the fact that each study examined should be clearly distinct from others. In practice, though, human nature leads to adaptation rather than revolution and the first reaction of any researcher is to look at the previous literature. Of more direct relevance, the possibility of research work being published and encapsulated in a meta-analysis tends to rise if it has strong links with an established literature. This is, of course, a problem in most fields of research, and is probably no worse in microeconomics than in other disciples, but again it can lead to results being discarded which fall excessively out of line with the mainstream. Tied to this, over time the conventional wisdom changes and this puts additional emphasis on the need to isolate 'time of analysis' in any statistical synthesis.

Further, in many fields such as microeconomics there are numerous consultancy studies undertaken, by both the public and private sectors, that are simply not accessible for meta-analysis and, hence, a body of valuable information is often inevitably excluded. And even when it is accessible, information retrieval systems often omit many of the studies. Problems of confidentiality are likely to be particularly acute in competitive type markets, when information can convey market power, and in cases where their release may lead to public concern. In fact, this is often the situation in areas of most interest to microeconomists. Even if these studies are let into the public domain findings are often presented selectively and are used as vehicles of advocacy rather then education. Such partial releases can be distortive. It is, therefore, good practice to set out explicitly the criteria for selecting the studies included in any meta-analysis and to indicate the limitations of the selection process.

In terms of the need for similarity in output measures, there are clearly areas where problems exist in the microeconomic literature as elsewhere. While studies of, for instance, demand-behaviour provide fairly standard outputs in terms of elasticities, and it is sometimes possible to manipulate results into such a common formulation even when there is no original standardisation, (e.g. valuations of externalities such as congestion, noise nuisance and accidents, are generally presented in specified ways) this cannot always be easily done. This tends to be particularly true where qualitative factors are involved and where diverse units are used. This is, of course, not to say that important developments are not being made in the more qualitative fields of economics which may make it easier in the future to embrace rather more qualitative and normative considerations within a meta-analysis framework. Further, efforts to circumvent problems associated with sets of studies of the latter kind by translating differing but normally continuous variables into common discrete variables can be distortive (Hunter and Schmidt, 1990).

4.4.2 Particular application problems in economic research

Many of the above arguments can be seen to relate to the more general problems of adopting meta-analysis reviews in the social sciences. Some of these are likely to be as prevalent if it is deployed in reviewing previous works on environmental externalities or labour efficiency as they are in any non-economic field of study. There are also, however, some rather specific difficulties which are linked with microeconomic research and the nature of its reporting.

First, perhaps the main underlying problem is related to the nature of most microeconomic research. With the exception of a limited number of recent embryonic studies, the microeconomic empirical method is one of quasi-experimentation rather than strict experimentation. It is still virtually impossible to conduct well controlled laboratory experiments in microeconomics, although the situation is now improving. Useful guides to experimental economics are Plott (1982) and Hey (1991). The problem is compounded by the qualitative nature of some aspects of economic research which constrains the use of many techniques commonly used in the natural sciences. The outcome has been considerable ingenuity on the part of economists in developing methodologies which can be employed but with this comes diversity and lack of standardisation. Meta-analysis becomes more vague in these conditions and the role of traditional reviewing procedures more important. These problems related to the qualitative nature of environmental policy problems are, however, largely the same types of problems which existed in the past in econometrics and related fields and which are now routinely handled in the work which is done. The problem may be quite simply that the efforts to date in applying meta-analysis to environmental policy assessment have focused on issues which can be viewed in qualitative terms for instance, exploring why studies produce quite wide variations in the values placed on traffic noise nuisance or the value of safety improvements. The quantification issue should not, therefore, be seen as a binding constraint.

Second, in the medical field, where meta-analysis is most widely deployed, there is considerably more standardisation in the way results are reported than in microeconomics. This makes its easier to pull together studies for comparison. Anyone who has looked at the literature in microeconomics discovers very rapidly that individual academic researchers have their own styles of presentation and that even wider variations exist when efforts are made to bring constancy findings into the analysis. Unlike strict experimental results, the reporting of assumptions, error distributions, data idiosyncrasies and so on is not standardised in economics. This point is made in a slightly different way by Goodwin (1992) when he discusses the differing nature of the dissemination cultures as between the academic, refereed journal route and the official or consultancy report channel. There are important differences in the pre-release evaluations they endure.

A somewhat different problem arises in the context of synthesising results from the increasing number of stated preference studies of microeconomic behaviour which are being conducted - for instance the use of contingent valuation methods in environmental economics. Unlike revealed preference analysis, where past actions are reviewed, stated preference analysis involves hypothetical, ex ante assessments often based on questionnaires and interviews. The implication for conducting meta-analysis is that examples not only involve variations in the way results are reported, but also their results relate to a particular set of unique questions and possible responses. These cases are difficult to handle although some efforts have been made in non-economic fields such as social psychology. These have only tended to be successful, though, where similar types of analytical procedures are adopted in each study.

Below we summarize the main existing limitations to and main criticisms on the application of meta-analysis in a more schematic way. These are respectively:

- Garbage in and garbage out is camouflaged by fancy statistics; this may apply certainly to the level of specific studies (Wachter, 1988).
- Elementary mistakes are made when a *potpourri* of study outcomes is treated with procedures invented for a statistical sample of experimental outcomes without the benefit of (i) controlled conditions, (ii) homogeneous measurement scales, and (iii) statistical independence that makes the procedure valid (Wachter, 1988).
- Doing a proper meta-analysis requires looking at all the studies done in the field, also the unpublished ones (*The Economist*, 1991). But here it is easy to miss out on some.
- The original studies might be biased (*The Economist*, 1991). However, this charge one can easily extend to other studies.
- There may be some arbitrariness in the selection of case studies (*The Economist*, 1991). The question is whether one should include all studies done in the field or lay criteria for selection down in advance.

Although the validity of some of these statements cannot be questioned, it ought to be recognized that such remarks apply to any comparative study. Meta-analysis tries to offer an appropriate framework to avoid such weaknesses to the maximum extent possible. In relation to these points, evidence is never given that studies with poor methodological design, measurement, treatment etc. have the handicap that these deficiencies influence their findings. The influence of study quality on findings has been regarded as an empirical *a posteriori* question, not an *a priori* matter of opinion or judgement used to exclude large numbers of studies from consideration (Glass *et al.*, 1984).

Altogether, these issues can be seen as major challenges to develop further the analytical tool of meta-analysis. This may be substantiated by offering examples from the economics literature in the next section.

4.5 Examples of the Use of Meta-analysis in Economic Research

4.5.1 Introduction

While the applications of meta-analysis to various areas of microeconomics are relatively small in number, there are some important areas where studies have been attempted. There is no claim though, that the material which is described below is comprehensive in its coverage since it is intended to provide a flavour of the type of work which has been undertaken rather than being a complete annotated bibliography of the relevant literature.

Looking over the literature, it is perhaps worth noting that the creeping interest in statistical synthesis exhibited by microeconomists often has its origins in issues related to management science or social psychology (e.g. its increased use in analysing various aspects of labour markets). There is also evidence that the number of meta-analytical studies has been increasing over time, possibly as people have become familiar with the potential of the approach, but also as the body of suitable studies for analysis has expanded and the demand for information has risen. Here particular attention is paid to environmental economics, labour economics, industrial efficiency and transport economics.

4.5.2 Environmental economics

Interest in environmental economics has expanded considerably in recent years. With this has come an increased attention to the need to integrate environmental costs within standard microeconomic assessment procedures. This has also been linked to something of a renaissance of interest in the use of traditional cost-benefit techniques based on expressing environmental effects in common monetary units. We may refer, for example, here to Cropper and Oates (1992) and Smith (1989). The result has been a considerable number of studies that have sought to place monetary values on negative environmental externalities. These studies have looked at a range of situations and have deployed a variety of different modelling and estimation procedures. Johansson (1987) and Freeman (1993) offer a detailed account of the methods of evaluation in common use. Pearce and Markandya (1989) offer a traditional review of the literature with, for instance, tables listing results from a diversity of studies on such things as values of life, air pollution and noise. The ranges of values obtained from different works are regularly cited but little other effort is made at statistically analysing the results. They also often explicitly recognise that in many cases they are measuring different things (e.g. the existence value of an amenity as opposed to a bequest value). Given the number of studies which have been completed, and the explicitly quantitative nature of the results which they have produced, there has been surprisingly little meta-analysis of their results. However, recently things are changing quickly, as shown in Table 4.2.

What has been attempted has focused on defining moderator variables and seeking out standard point estimates. There has been much reliance on regression procedures to seek out causal influences on result variations and, additionally a number of studies have attempted to produce simple summary statistics (e.g. means standard deviations and ranges) representing the general pattern of the findings attained (Smith and Kaoru, 1990a,b; Smith and Huang, 1993 and 1995; Button, 1995). Table 4.2 offers a brief summary of some of the main studies to date.

Table 4.2. Major studies in environmental, regional, urban and transport economics using meta-analysis.

Subject area	*Meta-analytical evaluation*
1. Urban pollution valuation	Smith (1989), Smith and Huang (1993), Smith and Huang (1995), Schwartz (1994), Chapter 10 (this book)
2. Recreation benefits	Smith and Kaoru (1990a), Walsh *et al.* (1989)
3. Recreational fishing	Sturtevant *et al.* (1995)
4. Valuation of life estimates	Chapter 11 (this book)
5. Contingent valuation versus revealed preference	Carson *et al.* (1996)
6. Noise nuisance	Nelson (1980), Button (1995), Chapter 4 (this book)
7. Congestion	Waters (1993), Button and Kerr (1996)
8. Internal validity of contingent valuation and visibility improvement	Smith and Osborne (1996)
9. Multiplier effects of tourism	Chapter 9 (this book)
10. Transport issues	Chapters 13 and 14 (this book)
11. Price elasticity of demand in travel cost method studies	Smith and Kaoru (1990b)

To provide an indication of how meta-regression analysis can be deployed in this field, Table 4.3 offers a list of some of twenty five studies which have been conducted on the effects of noise nuisance on property values (expressed in terms of percentage reductions in house values as a result of one additional unit of noise nuisance).

The table offers information concerning the methodologies used in the studies, the countries to which they relate, the sources of noise nuisance and the data types involved. Linear ordinary least squares, meta-regression produces the results set out in Equation (4.4). Since the studies used in this type of work involve different sample sizes and independent variables, it is also normal to correct for heteroskedasticity but, since these calculations are meant to be merely illustrative, this is not done here. The general form of this equation is Equation (4.1).

$$\text{Value of Noise Nuisance} = 0.985 - 0.499\text{Australia} - 0.549\text{UK} - 0.289\text{USA} - 0.318\text{Disaggregate} + 0.282\text{Airport} \quad (4.4)$$

The model explains 40 percent in the variations in noise nuisance values which have been reported in the studies surveyed. The country dummy variables, although not all highly significant, show the importance of the country to which the studies relate - Canada, the base nation, seemingly being more highly sensitive to noise levels than others in which work is reported. Equally, the noise nuisance associated with airports seems to generate more of a problem than comparable noise from highways; although the airport variable only exhibits a 88 percent confidence interval.

Table 4.3. Summary of studies examining the money value of noise nuisance (expressed as percentage of house values).

Study	*Method*	*Value*	*Country*	*Data type*	*Noise source*
Abelson, 1979	Hedonic Prices	0.4–0.5%	Australia	Disaggregate	Airport
Anderson and Wise, 1977	Hedonic Prices	0.25%	USA	Disaggregate	Highway
Bailey *et al.*, 1963	Hedonic Prices	0.3%	USA	Disaggregate	Highway
Collins and Evans, 1994	Neural Networks	0.3–0.6%	UK	Disaggregate	Airport
De Vany, 1974	Hedonic Prices	0.8%	USA	Aggregate	Airport
Dygert, 1973	Hedonic Prices	0.5–0.7%	USA	Aggregate	Airport
Emerson, 1969	Hedonic Prices	0.57%	USA	Disaggregate	Airport
Gamble, 1974	Hedonic Prices	0.3%	USA	Disaggregate	Highway
Gamble, 1974	Hedonic Prices	0.26%	USA	Disaggregate	Highway
Gautrin, 1975	Hedonic Prices	0.35%	UK	Disaggregate	Airport
Hall *et al.*, 1978	Hedonic Prices	1.05%	Canada	Disaggregate	Highway
Langley, 1976	Hedonic Prices	0.32–0.4%	USA	Disaggregate	Highway
Levesque, 1994	Hedonic Prices	1.30%	Canada	Disaggregate	Airport
Mcmillan *et al.*, 1978	Hedonic Prices	0.50%	Canada	Disaggregate	Airport
Mcmillan *et al.*, 1980	Hedonic Prices	0.13–1.6%	Canada	Disaggregate	Airport
Maser *et al.*, 1977	Hedonic Prices	0.55–0.95%	USA	Disaggregate	General
Miezkoski and Saper, 1978	Hedonic Prices	0.32–0.4%	Canada	Disaggregate	Airport
Nelson, 1978	Hedonic Prices	1.10%	USA	Aggregate	Airport
Nelson, 1978	Hedonic Prices	0.8%	USA	Aggregate	Highway
O'Bryne *et al.*, 1985	Hedonic Prices	0.52%	USA	Aggregate	Airport
O'Bryne *et al.*, 1985	Hedonic Prices	0.57%	USA	Disaggregate	Airport
Paik, 1972	Hedonic Prices	2.2%	USA	Aggregate	General
Price, 1974	Hedonic Prices	0.83%	USA	Aggregate	Airport
Uyeno, 1993	Hedonic Prices	0.65–1.6%	Canada	Disaggregate	Airport
Vaughan and Huckins, 1975	Hedonic Prices	0.6%	USA	Disaggregate	Highway

Not surprisingly perhaps, given the nature of the case studies available, work of this kind, seeking moderator variables, has tended to have been conducted on particular forms of environmental damage rather than on overall environmental impacts.

Nelson (1980) similarly uses regression techniques to examine the adverse effects of airports on local noise pollution and, from 13 studies, produces a discount range of 0.4 to 1.1 per cent with a mean of 0.62 per cent. A degree of subjectivity is introduced into this work, however, since the results are based 'on what, in the author's opinion, are the best coefficient estimates obtained in each study.' Recently, another study has considered a similar problem (Schipper, 1996).

Because of its diverse nature and the difficulty of gaining accurate scientific information, studies concerning atmospheric pollution are limited but, nevertheless, some meta-analysis of their results has been attempted. Smith (1989), for example, considers empirical work looking at 35 studies of local air quality where monetary values were derived using hedonic property value models. Again consistency in the results emerges once allowance is made for local conditions and the assumptions made. Smith's (1989) conclusion on work on evaluation of external values is that, "Meta-analysis of results over a wide range of studies indicates that modelling assumptions matter (as we would expect) but that estimates are not pure noise. They are systematically related to the features of the resource and to the very assumptions most analysts would argue should matter." Schwartz (1994) also conducted a meta-analysis reviewing work linking atmospheric pollution to morbidity.

More recently, Smith and Huang (1995) have examined hedonic price studies conducted between 1967 and 1988 on the marginal willingness to pay for reducing particulate matter in the air. Some 167 published and unpublished hedonic models were examined although, because of the nature of many of these, and in particular lack of information regarding factors such as city characteristics, the meta-analysis was forced to rely on 86. In many cases, the studies used in the meta-analysis had to be reconstructed to obtain estimates of target variables such as the marginal willingness to pay. Since a number of variations of methodology had to be used to develop this data base, this may have influenced the findings to some extent. Further, in many instances a number of estimates were taken from individual studies (up to 20) which reduces the strict study independence assumptions of meta-analysis. Ordinary least squares with a Huber consistent covariance matrix and minimum absolute deviation estimators were used for estimation purpose. These procedures were selected to reduce the particular problems of outliers deviation and heteroscedasticity. Neither, however, deals with the inherent problems for which the other is designed. The results of the meta-analysis show that market conditions, and the procedures used to implement the hedonic models, were important in explaining variations in the values individual studies produced (for more details and an alternative analysis see Chapter 10). The actual numerical results examined involved an interquartile range for estimated marginal values (changes in asset prices) which in the studies examined was between zero and $98.52 (1982-84 prices) for a one unit reduction in total suspended particulates (in micrograms per cubic meter). The mean marginal willingness to pay was nearly five times the median. Outlier findings, therefore, existed across the studies.

The question of the value of recreational amenities which would be lost if land were transferred to other uses has proved a fertile ground for environmental valuation studies in recent years. In particular, there have been a significant number employing the recreation demand method originally proposed by Hotelling, often called the travel cost method. In operational terms this involves using indirect revealed preference theory and looking at how much individuals pay to travel to an unpriced recreational facility and, from this, derive the demand curve, and ipso facto, the consumer surplus associated with the existence of the amenity.

Work by Smith and Kaoru (1990a; 1990b) looked at some 200 published and unpublished studies of recreational demand appearing between 1970 and 1986 concerning the benefits derived from recreational resources which employed the travel cost methodology which stems from Hotelling's idea. They used regression based meta-analysis, and explored a number of equation specifications, to examine the results of 77 of these studies. Twenty-one variables were included in the most comprehensive specification. It was found, once again, that, while the travel cost approach has a degree of official acceptance in policy formulation, results are very sensitive to the underlying modelling assumptions. In particular, modelling decisions in this field are often data-based rather than founded in economic theory and this was an important influence on the estimated price elasticities across studies. Equally, the meta-analysis gave support to issues which the wider literature had suggested to be important in deploying the travel cost method. One of the main conclusions from looking at this body of analysis, therefore, was support for earlier work by Walsh *et al.* (1989), namely that variations across studies can often largely be explained in terms of the specific nature of the recreational resources and the underlying assumptions made in the estimation models employed.

An important question with respect to the choice of an economic valuation method is whether one method leads generally to upward or downward biased estimates of monetary values of environmental amenities. Meta-analysis seems particularly useful to address this issue, as it allows for a pairwise comparison of methods, often using large samples of past studies. At a general level it is often possible, by pooling methods, to construct a large meta-analysis sample by including valuation estimates for various types of environmental goods or bads. Carson *et al.* (1996), for example, used 83 studies to render 616 comparisons of contingent valuation (CV) and revealed preference (RP) estimates, with the latter based on applications of travel cost techniques, hedonic pricing techniques, expenditure and household production function models, and simulated or actual market creation. The estimates cover a variety of environmental issues that fall in the class of quasi-public goods, and include recreation benefits, water quality improvements, fishing, mining impacts, forest harvesting, preservation, and job-related risk reduction. These were coded into three broad classes named recreation, environmental amenities (air and water quality) and health risks. It is found that CV estimates are generally, though not always, smaller but only slightly smaller than their RP counterparts. In other words, this provides support for convergent validity of the two basic approaches to non-market valuation. The same conclusion is drawn even if both general type of methods (CV and RP) cannot avoid errors in value estimates. This is supported by all the statistical tests applied to CV/RP ratios, RP/CV ratios, (CV-RP) difference, and vote-counting and for all of the following subsets: the complete dataset; a 5 percent trimmed dataset (to ameliorate the large influence of gross outliers); and a weighted dataset giving equal weight to each case study rather than each single CV/RP comparison within case studies, effectively using the mean CV/RP ratio for each study as that study's observation. Although each approach leads to unique statistical indicators, the qualitative conclusions are identical.

Recently there has been much debate on whether contingent valuation can be relied on to generate indicators for values which its proponents claim to estimate. Since the Exxon-Valdez oil spill and the subsequent evaluation of the contingent valuation method by the NOAA panel (the General Counsel of the National Oceanic and Atmospheric Administration; Arrow *et al.*, 1993), a number of respected economists outside the circle of environmental economics have focused their attention on this method (see Hausman, 1993). Their main concern is that when nonuse values dominate, willingness to pay (WTP) functions do not necessarily comply with properties that economic theory implies. An internal validity test is therefore necessary. The NOAA panel proposed to perform a scope test, requiring the contingent valuation WTP estimates to be responsive to the amount that is available of the environmental amenity. Smith and Osborne (1996) perform such a scope test based on a meta-analysis of past case studies estimating the marginal willingness to pay for (improved or maintained) visibility in natural parks (indicator in distance (km)). The idea underlying their approach is that in order to trace how the WTP reacts to various changes in relevant observable factors, sufficient variation of observations is required. The (second-)best approach is then to use past case study estimates, i.e. a meta-analysis sample. Using estimated relationships of five main studies, together rendering between 88 and 115 meta-analysis observations, various models were estimated, based on leaving out specific observations (case studies), and dealing with a number of specifications (semilog and linear) and different sets of independent variables. In all meta-analysis estimations a statistically significant, positive relationship is found to hold between WTP and the proportionate improvement in the visible range. For a specific quasi-linear, indirect utility function, Diamond (1996) derived that a scope test would require the ratio of two WTP estimates for two different physical indicators (e.g. numbers of birds saved) to be greater than the ratio of these physical indicators. The meta-analysis results are consistent with this requirement for all different relevant models investigated. In other words, for this particular issue the contingent valuation method passes the scope test, and as argued by Smith and Osborne (ibid.), not merely on the basis of statistical significance but also on the basis of economic plausibility.

Many meta-analysis studies focus on nonparametric density estimates for the meta-analysis dataset, or use statistical correlation measures, to examine if estimates are significantly different between studies, estimated relationships or techniques. Other studies have used regression methods to understand the factors causing differences between estimates among studies or estimated relationships. These factors may include case characteristics (e.g., according to resource, region, time), study characteristics (e.g., method used, sample size) or meta-characteristics (e.g., published study or not). Sturtevant *et al.* (1995) argue that some meta-analysis datasets can often be regarded as panel data, and illustrate the usefulness of using panel estimation techniques in the context of fishery recreational demand studies. The underlying idea is that regression based on ordinary least squares does not deal adequately with the correlated error structure of the data and that modified estimation procedures are, therefore, needed. Correlation occurs among multiple estimates from single studies, resulting from the use of subsets of the complete dataset available, different model specifications, different indicators or other assumptions.

In order to deal with systematic effects on estimates obtained within single studies, panel models are suggested. Sturtevant *et al.* report the results of three such models, namely fixed-effects, random-effects and separate variances' models. Panel data are treated based on distinguishing groups, and in the context of a meta-analysis groups can be defined based on the criterion that estimates belong to a single case study and possibly other criteria, such as type of problem or type of resource. The various panel models can deal with different phenomena, e.g. the intercept or the error term varying across groups (i.e. studies). The best fit panel model has significant estimates for all parameters that are significantly estimated by an OLS model. Other interesting points are that the best fit panel model leads to the lowest predictions of all the models estimated, and that variation among groups is required to be able to predict for new cases. Predictions for groups not included in the panel are not, however, possible.

One of the uses of meta-analysis is for benefit transfer, notably in cases where time or money for indepth studies is limited, or when application of results obtained for one case to another seems evident. Using the terminology of Sturtevant *et al.* (1995), basic benefits transfer can be performed, meaning that the estimate from the best study extracted from the relevant literature can be transferred to the new case under consideration. Instead, one can take the mean of the estimates extracted from case studies that have been conducted on similar types of problem. Alternatively, one can use a regression model obtained from the closest study to predict for other cases using appropriate values for the variables in the model. Finally, a meta-analysis model (or multiple models) can be used to predict. Comparisons of such predictions can, among others, be found in Sturtevant *et al.* (1995) and Smith and Huang (1995). Both find that basic transfer, based on mean values obtained from similar studies, leads to higher estimates than various meta-analysis models.

4.5.3 Labour economics

Given the interface between microeconomics and management science, and the greater use made of meta-analysis in the latter, it is perhaps not surprising that topics on the margin of the two subject areas have been the basis of more meta-analysis than is common in many other areas of microeconomics. In particular, a number of meta-analytical studies have been done looking at various aspects of the labour market. There have, for instance, been meta-analyses of labour market responses to market stimuli, e.g. the study by Fiset *et al.* (1994) of dentists; to the importance of education, e.g. Phillips (1994) on farmers' education; to union membership (Jarrell and Stanley, 1990); and, slightly more remote, to the importance of profit sharing on productivity (Weitzman and Kruse, 1990).

The largest group of studies in this area, however, has been concerned with absenteeism, and specifically its possible link with turnover. Much of the interest stems from a pioneering meta-analysis by Hunter *et al.* (1982). Using bivariate analysis they found a positive correlation between absenteeism and turnover.

Up-dating of this work has involved the analysis of 24 studies of absenteeism by Mitra *et al.* (1992), although, because of problems such as the common use of data sets and difficulties in converting results into a standard format, only 17 studies, embracing 33 absenteeism correlations, were statistically examined. The statistical technique adopted by Mitra *et al.* generally ignored the possible violation of independence assumptions and his approach involved looking at covariation while focusing on moderators such as the type of absence measure used, the duration of studies and industry type. The data base for this meta-analysis consisted of studies found from a computer bibliographic data-base sweep, a manual search of key journals and a reference search of seminal papers in the field. A search for unpublished data (notably dissertations) was also conducted but the results were of no practical use. This degree of seeking out important works for meta-analysis must be seen as more thorough than is the norm in the social sciences. The analysis found that nearly 73 per cent of the observed variance in correlations across studies could be explained by the study artefacts of sampling and measurement error after correcting for errors due to dichotomising the absenteeism variable and unequal sample sizes. The more important microeconomic result was the detection of a positive correlation between absenteeism and turnover indicative of a continuum of withdrawal behaviour progressing from absenteeism to turnover. This result is also consistent with two meta-analysis of the relationship between job attitudes and absenteeism (Hackett and Guion, 1985; Scott and Taylor, 1985) and a meta-analysis of job attitudes and turnover (Steel and Ovaille, 1984). Mitra *et al.* also found, but could not adequately explain, that the studies explored suggest there is a higher probability of this positive correlation occurring in manufacturing than non-manufacturing sectors.

Hackett (1990) was more interested in the nature of the workers involved. Using the Hunter *et al.* framework, and applying their 75 per cent cut-off rule to isolate moderator variables (i.e. when 75 per cent of the observed variance in relational coefficients have been explained), he found that age, but not tenure, was inversely related to avoidable absenteeism.

Links between absenteeism and labour performance were explored in a meta-analysis of 46 correlation studies by Bycic (1992). The findings are that there is a modest, but significant, relationship between frequently absent employees and poor performance. The variations between studies could not, however, be explained in terms of differing statistical artefacts. In a similar vein, Williams and Livingstone (1994) conducted a meta-analysis of 55 studies looking at labour performance and voluntary turnover. A negative relationship is found which is invariant to the unemployment rate.

4.5.4 Industrial efficiency

Recently there has been an up-surge of interest in the effectiveness of economic regulations and, in particular, whether they might prove detrimental to X-efficiency. Recent advances in both econometrics and programming have provided analysts with tools to explore the extent to which X-efficiency exists. X-efficiency is theoretically different to technical efficiency but techniques such as stochastic production function fitting and data envelope analysis cannot separate the two and hence either may actually be measured. Button and Weyman-Jones (1992; 1994), in a small scale meta-analysis employing rank correlation procedures, looked at studies of 9 US and European financial institutions and government and other agencies which had employed data envelope analysis (DEA) to assess levels of X-efficiency. The sample of studies was itself initially vetted and a selection made from a larger pool with those with less than 35 degrees of freedom being rejected.

As seen in Table 4.4, the analysis found a positive rank correlation between the ranking of institutions in terms of degree of X-inefficiency and the level of bureaucracy (constraint concern pressure) affecting them. Also, privately owned, competitive or weakly regulated firms have a narrower dispersion of inefficiency. A clear limitation of the work, however, was the normative nature of the definition of constraint concern which, while embracing easily measured features such as private/public ownership differences also rested on judgements regarding regulatory intensity.

Table 4.4. Rank correlations between efficiency measures and the index of lack of constraint concern.

Mean efficiency	Standard deviation	Minimum efficiency
-0.18	0.57	-0.39

Efficient industrial location choices, in the light of the changing composition of the industrial base and concern over regional depravation problems, have been the subject of a considerable number of recent empirical studies. The implications of transport investment for the local economy in terms of development impacts has long been contentious. Neither theory nor empirical analysis provides conclusive evidence of the key interactions. In particular, and perhaps partially representing a microeconomic reflection of the macroeconomic debates about the relevance of infrastructure provision for economic growth, the question of the implications of transport investment on firms' location and local development has been regularly examined.

In addition to a considerable literature on the link between transport and economic development, there are a growing number of studies concerned with the effects of taxation and similar fiscal incentives on micro- and mesoeconomic performance. Phillips and Goss (1995) examined 60 US inter-metropolitan studies and 9 intra-metropolitan studies which have estimated the long-run elasticity of business activity with respect to state and local taxes. This regression work follows on from a much simpler analysis of a fuller data set of 84 econometric studies by Bartik (1991) which limited itself to simply averaging elasticities for part of his data base. Efforts to control for factors other than inter/intra metropolitan differences were very limited. The most complete set of regression results they estimated, based on the former subset of data, is replicated in Equation (4.5).

$$\begin{aligned}\text{Tax Elasticity} = {} & -0.346 - 0.145\text{Fixed} - 0.255\text{Public Sector} + 0.193\text{Wage} \\ & + 0.026\text{Single} + 0.185\text{Firm} + 0.240\text{Capital} - 0.166\text{Welfare} \\ & - 0.100\text{Manufacture} - 0.010\text{Business Tax} - 0.078\text{Education} \\ & + 0.058\text{Energy} + 0.1101\text{Climate} + 0.010\text{Union} \\ & - 0.079\text{Population Density} \qquad (4.5)\end{aligned}$$

The explanatory power of the meta-regression model is low ($R^2 = 0.253$) and many of the variables are statistically insignificant. This poor model fit is in line with that of many other meta-analysis that have deployed regression analysis and may reflect the need to throw more high powered technique at the topic or to think through the appropriate variables to include more carefully. What emerges is that considerations such as whether individual studies controlled for fixed effects and/or public services, partly determine the size of the elasticities that they generated. In general, though, the differences in methods of analysis do not lead to different estimates of tax elasticity. However, studies that failed to control for public service effects tended to underestimate the tax elasticity while studies that measure growth as aggregate income or investment growth will produce a lower tax elasticity.

4.5.5 Transport economics

Major transport infrastructure investments often have considerable microeconomic impacts on local communities. They have both immediate implications for travel behaviour and, as has been examined in the previous subsection, may interact with the wider economic system to influence industrial location and the spatial distribution of employment. These implications have provided the basis for considerable empirical analysis over the years (Winston, 1985). The number of studies, in part, reflects the nature of institutional arrangements in many countries where local evaluations and public inquiries form part of the transport assessment and planning process. Such studies are often a statutory requirement. Efforts to statistically bring together the findings of such works on traffic matters, mainly because methodologies are often not standardised, have until recently been relatively limited.

In terms of strict direct traffic and transport effects, Waters (1993) explored 56 wide-ranging, international studies to seek out commonality in the monetary values placed on the travel time savings associated with improved transport. The conclusion was that it amounted to about 30 to 50 per cent of the wage rate but with a large standard deviation. Extending the data set, and looking at the effects of the country of study, the time of the study and trip purpose (see Equation 4.6), Waters (1995) subsequently found some significant explanation for the variability in the values of travel time savings in terms of the importance of an interurban effects and a time of study variable. Overall, though, the factors examined provide only a poor explanation of the diversity of estimates found in money value of travel time studies - the R^2 value, for example, was 0.191. Waters could not categorise sufficient studies to explore the importance of the modelling techniques used in the travel time savings work although this has often been held to be an important consideration in the transport economics literature and has been found of importance in meta-studies of other microeconomic topics.

$$\begin{aligned}\text{Value of Travel Time Saving} = {} & 0.447 + 0.008\text{Year} - 0.053\text{USA} - 0.082\text{UK} \\ & - 0.202\text{Australia} - 0.029\text{Commuting} \\ & + 0.0203\text{Interurban} \qquad (4.6)\end{aligned}$$

An equally extensive number of empirical studies have been conducted on the demand elasticities for transport. Goodwin (1992) provides a recent meta-analysis which seeks to determine summary statistics (point estimates) and offers some simple guidance (ranges) of the dispersion. This work can usefully be compared to a similar study by Oum *et al.* (1992) which is much more in the traditional literary mould with tabulations and listings of results together with a verbal assessment of why variations occur. The work covers the price elasticity of demand for petrol, of traffic levels with respect to fuel costs and bus transport with respect to fares. A range of findings emerge, of which perhaps the most important from a policy perspective is that in an analysis of 50 studies of public transport demand with respect to fares, the elasticity which emerges is considerably higher than previous, more traditional, surveys had suggested. It is, however, taken as axiomatic that there will be differences according to whether time series or cross-sectional analysis is performed but other than separating the results this is not examined rigorously.

The implications of transport investment for the local economy in terms of development impacts has long been contentious. Neither theory nor empirical analysis provides conclusive evidence of the key interactions. Conventional examination of the individual studies which have been conducted and setting out their results in tabular form offers little guidance with some revealing a positive correlation, other a neutral effect and some a negative impact.

4.6 Conclusions

An extensive variety of meta-analysis techniques are now widely and regularly deployed in the natural and medical sciences. One strong finding from various meta-analyses in these fields is that most new treatments have, at best, small to modest effects. Gilbert *et al.* (1975) found this for their summary of three dozen social and medical programmes. Few of these innovations showed large, positive effects with only a modest number showed any significant positive effects. Light (1983) presents the result of two dozen meta-analysis that are exemplary and finds a similar result. He concludes that most innovations have at best small positive effects that such small positive findings should be treasured rather than ignored and that the importance of this finding is that managers of programmes should understand they should not expect large positive findings to emerge routinely from new programmes. Most meta-analysis in natural and medical sciences have concentrated on assessing whether a given type of intervention has a particular type of effect. The better meta-analyses have also explored some of the method factors, some of the populations and settings and some of the treatment variants that influence the size of the effect.

The use of meta-analysis in applied, quantitative microeconomics is much less common. In some fields, such as labour economics, they are gradually being adopted more liberally, possibly because of the interface with areas where the technique has a longer history, in this instance with management science. Environmental economics is also gradually developing experience with meta-analysis although here the stimulus has more to do with seeking consensus on methodology and approach to particular issues, than to its links to the harder sciences. In all cases, the body of additional knowledge which is emerging is not trivial although, equally, it is not always earth shattering.

The current approaches to meta-analysis, and in particular the techniques used in the microeconomic field, however, still have a number of limitations associated with them. In particular, the raw material they have to work with is often biased, not only because of its reliance on published sources but also because of the lack of standardisation in methodology and reporting. The qualitative nature of much of the work is also less easily handled using the standard statistical tools of meta-analysis.

While there are these problems there are also indications that on-going changes within economics and in statistical synthesis analysis may facilitate both more meta-analysis in the future and the adoption of the techniques across a wider range of sub-fields. One important trend is that with the widespread adoption of 'off-the-shelf' computer software there is increasing standardisation of the ways individual case studies are conducted and the results presented. Additionally, the considerable advances in information retrieval systems means that it is easier for those conducting meta-analysis to carry out searches and ensure the widest possible set of studies are embraced in their works. As these forces become more important so there is also an inevitable learning-by-doing process with economies of experience improving the quality of the meta-analyses which are conducted. These economies may, for instance, be in terms of the techniques to use but could also embrace less bias selections of studies to include.

Perhaps of equal importance are advances in statistical procedures which are increasingly allowing qualitative factors to be embraced in microeconomic analysis. This is particularly relevant in many areas of micro-economics where quantification is often not perfect or, in many instances, not possible. Reflecting on the microeconomic meta-analysis completed to date, one is struck by the heavy reliance on regression methods and the widespread use of dummy variables to capture many effects. One may anticipate in the future that more sophisticated parametric procedures could be used for analytical puposes, for instance in relation to qualitative considerations (Amemiya, 1981).

There are also developments in areas outside of the conventional domains of econometrics which may provide more useful ways forward. Non-parametric techniques, and especially various forms of soft-modelling, would seem to offer a viable way of handling some of the problems which are often inherent in the types of techniques which have been used until now. One possible solution to problems of comparing differing measuring procedures and those associated with making distributional assumptions, for instance, could be the adoption of rough and fuzzy set analysis (see Chapters 7 and 8). Developments here should provide the scope for further work in more policy oriented areas which often involve considerations outside of the narrow confines of monetary measures (e.g. the effectiveness of alternative policies on the quality of life). These developments are important if the applications of meta-analysis are to provide a balanced output. There is a danger, for instance, that its application to areas where immediate quantification is possible, such as price elasticities, could lead to a bias in thinking regarding the importance of less immediately quantifiable factors such as taste.

In conclusion, it is clear that meta-analysis has a useful role to play in the study of economics and environmental economics in particular. The indications are that there is the potential for it to be further tailored to the particular needs of those working in this field. It should not, however, as some commentators have suggested, be seen as a superior substitute for conventional reviews but rather as a very powerful additional way of gaining insights from a body of existing studies.

PART B

METHODOLOGY OF META-ANALYSIS IN ENVIRONMENTAL ECONOMICS

CHAPTER 5 STATISTICAL META-ANALYSIS

5.1 Introduction

In the previous chapters the general concept of meta-analysis was discussed. This chapter is devoted to one particular group of statistical techniques which can be used for a meta-analysis. These are the techniques first used as meta-analytical tools in medical and social sciences.

The aim of this chapter is to present a technical background of meta-analytical statistical methods. A general and concise review will be given of the most relevant issues related to statistical meta-analytical methods. The review is mainly based on Hedges and Olkin (1985) and Cooper and Hedges (1994). Other useful guides on relevant statistical techniques include: Light and Pillemer (1984), Wachter (1988), Wolf (1986), Cook and Leviton (1980), Fiske (1983), Hunter *et al.* (1982), Rosenthal (1991) and Cook (1992).

Statistical methods for meta-analysis have been developed in order to combine the results of individual studies, and to analyze the variation of these. Such methods were designed for integrating results of medical and social studies. We start this chapter with some remarks about the general structure of medical research and the kind of information the statistical methods were designed to analyze.

In contrast with economic research, medical research is carried out in a quite uniform way. In addition, research reports also have a rather uniform structure. Medical research deals with the effectiveness of specific medicines and treatments to cure diseases. A simple medical or social-psychological experiment could have the following design. A group of people each suffering from the same disease, is divided into two smaller groups. Members of the first group are given the medicine or undergo a specific treatment, while the members of the second group do not receive anything. After a predetermined period of time, the effect of the medicine or treatment is measured with respect to some indicator. On the basis of these measurements an index of effect magnitude is estimated which is essentially equal to the difference of the values of the indicator before and after the treatment. Section 5.2 discusses the estimation of such indices of effect magnitude in more detail.

Most medical reports mention at least two characteristics of the results of the research (Rosenthal, 1991). First, they usually report the estimate of the size of the relationship between the dependent and the independent variable (the effect size of the medicine or the treatment). In the medical experiment described above, the dependent variable is the specific indicator which has to be measured; the independent variable is a bivariate variable, say X, with $X=1$ for the people who are under treatment, and $X=0$ for the people who belong to the control group.

In the second place medical reports usually mention an estimate of the accuracy of the effect size. They can do this by means of reporting the estimated variance of the estimated effect size. Another way of informing about accuracy is via a confidence interval of the effect size. In addition, reports usually mention whether the effect size differs significantly from 0 for a prespecified significance level.

The meta-analytical statistical methods that will be discussed in this chapter, use this kind of information, e.g. estimates of the effect size and estimates of their accuracy, of individual medical research studies. They can be divided into four categories according to the type of question they can address and the information which is required for their application. In the first place there are methods which are based on counting studies which have a significant result (i.e. the effect size differs significantly from 0 for a certain significance level). Assume that there is a number of independent studies available each of which which focus on a similar phenomenon. In each study one estimate of the effect size is calculated. Often it is assumed that the effect sizes found by the different studies are estimates of the same real effect size, the common effect size. Using the first type of method it is possible to construct a confidence interval for this assumed common effect size even if the values of the estimates of the individual effect sizes are unknown. Section 5.3 pays attention to this vote-count procedure.

The second group of methods consists of non-parametric methods for combining significance levels. The so-called test of combined significance tests the null hypothesis which states that the effect size is 0 in all studies that are included in the meta-analysis. This test will be discussed in Section 5.4.

Next, there are methods for combining the individual estimates of the effect sizes. These methods combine several estimates into one single estimate of the supposed common effect size. Following this, a confidence interval for this common effect can be constructed. Section 5.5 deals with these methods.

The question arises whether it is justified to assume that there exists a common effect size. The homogeneity test checks this assumption. Rejecting this assumption raises the question what the reason is for the variation in the found effect sizes. The availability of more than one study allows for an answer to this question: the influences of the surroundings and the way of research on the found effect sizes can be examined. The final group of methods consists of methods for answering such questions. They are discussed in Section 5.6.

5.2 Estimation of Individual Effect Sizes

Many medical examinations try to answer the question whether a specific medicine cures a patient of a particular disease. Which indicator is chosen to measure the effect of the medicine depends on the design of the examination. In the example in which one group is given the medicine and the other is not, the standardized mean difference is an appropriate measure. On the other hand, if each person is given a different dose of the medicine, the standardized mean difference is not an appropriate measure of the effect size. If the examination has this design, another statistic is an appropriate measure: the correlation between the dose and the observed effect. The following example illustrates the use of statistical methods in a specific situation.

Example

A group of n persons is divided into two smaller groups. The n_E members of the first group are given a specific medicine, or a specific medical treatment, while the n_C members of the second group, the control group, do not receive anything. After the cure or treatment a specific indicator is measured for each person, for the people belonging to the first group as well as for those belonging to the second. Let $Y_1^E, ..., Y_{nE}^E$ be the observations of the first group, and $Y_1^C, ..., Y_{nC}^C$ the observations of the second group. These data can be analyzed statistically.

It is assumed that there is a real fixed value of the specific indicator for each of the two groups. These real values are represented by the parameters μ_E and μ_C. The question is now what the effect size of the medicine has been. This effect size μ_E-μ_C, or the effect size standardized by the standard error of the observations,

$$\delta = \frac{\mu_E - \mu_C}{\sigma}, \tag{5.1}$$

can be estimated by an estimator D which is a function of the observations. If the observations are assumed to be generated by a probability distribution, D also has a probability distribution. On the basis of this probability distribution a confidence interval for δ can be constructed. It is also possible to test hypotheses about values of δ.

The estimate t of the effect size θ which is mentioned in a research report, is assumed to come from the probability distribution of the estimator T of θ. While estimating, an estimation error e is made (Cooper and Hedges, 1994):

$$e = \theta - t \tag{5.2}$$

e comes from the probability distribution of the estimation error E. The variance of E is the sum of the variance of the estimator T and the variance of the parameter θ:

$$\text{var}(E) = \text{var}(\theta) + \text{var}(T) \tag{5.3}$$

There are two approaches to var(E): the fixed effects model and the random effects model. The fixed effects model, supposes that there exists one fixed, real effect size, in other words: $\text{var}(\theta)=0$ and $\text{var}(E)=\text{var}(T)$. Reasoning from the fixed effects model, the difference between two estimates of effect sizes which have been estimated in the same way, is explained by the fact that each study estimates the effect θ for a different sample of the total population: the medicine was given to different groups of people.

Example (continued)

It is often assumed that the observations come from a normal distribution (Hedges and Olkin, 1985). Let $Y_1^E, ..., Y_{nE}^E$ be independent and normally distributed with mean μ_E and variance σ^2. Let Y_1^C, ..., Y_{nC}^C also be independent and normally distributed, but with mean μ_C and variance σ^2. In this example, the observations of the members of the control group and the observations of the members of the experimental group have the same variance.

A possible estimator of δ is the standardized difference of the sample means of the two groups:

$$G = \frac{\overline{Y^E} - \overline{Y^C}}{S^*}, \text{ with } S^* = \left(\frac{\sum_{i=1}^{n_E} (Y_i^E - \overline{Y^E})^2 + \sum_{i=1}^{n_C} (Y_i^C - \overline{Y^C})^2}{n_E + n_C - 2} \right)^{1/2} \tag{5.4}$$

G is a biased estimator of δ:

$$E[G] = J(n_E + n_C - 2)\delta \approx \left(1 + \frac{3}{4(n_E + n_C) - 9} \right)\delta \tag{5.5}$$

whit J being some transformation function. The estimator D, that has been derived from G, is an unbiased estimator of δ:

$$D = J(n_E + n_C - 2)\ G \tag{5.6}$$

D has also a smaller variance than G. D and G have both a probability distribution function with a very complicated analytical form. If n_E and n_C are both larger than 10, the probability distribution of D can be approximated by a normal distribution very well. This large sample distribution of D is given by proposition 1 found in Appendix 5.1.

The second approach, the random effects model, imputes differences between estimates of the effect size only partially to the composition of the sample. The idea is that it is impossible to carry out two examinations under exactly the same circumstances, for there are very many factors which influence the magnitude of θ, but which the researcher does not know. This ignorance leads to a stochastic real effect size θ: var(θ)≠ 0 and var(E) = var(θ)+var(T).

Example (continued)

The estimator D is an unbiased estimator of δ in case of the random effects model (Hedges and Olkin, 1985). However, the variance of D is larger than if δ had been a fixed effect. The real effect size δ comes from the probability distribution of the random variable Δ. The variance of D is equal to the sum of the variance of Δ and the conditional variance of D given δ:

$$\text{var}(D) = \sigma^2(\Delta) + \sigma^2(D|\delta) \tag{5.7}$$

The conditional variance $\sigma^2(D|\delta)$ is equal to the variance of D in the fixed effects model.

The variance of the real effect size, $\sigma^2(\Delta)$, can only be estimated if the effect size has been estimated more than one time; in other words, it is necessary that more than one study has been carried out. Let $d_1, ..., d_k$ be the estimates of the real effect sizes $\delta_1, ..., \delta_k$; the outcomes d_i, i=1, ..., k, are influenced by the probability distribution of the δ_i's and the composition of the sample. The estimator of the unconditional variance, $s^2(d)$, of these outcomes is:

$$s^2(d) = \sum_{i=1}^{k} \frac{(d_i - \bar{d})^2}{k-1} \tag{5.8}$$

Let $s^2(d_i|\delta_i)$ be the estimated conditional variance for i=1, ..., k. The mean of $s^2(d)$ is equal to:

$$E[s^2(d)] = \sigma^2(\Delta) + \sum_{i=1}^{k} \frac{\sigma^2(d_i|\delta_i)}{k} \tag{5.9}$$

An unbiased estimator of $\sigma^2(\Delta)$ is now:

$$s^2(\Delta) = \sum_{i=1}^{k} \frac{(d_i - \bar{d})^2}{k-1} - \sum_{i=1}^{k} \frac{s^2(d_i|\delta_i)}{k} \tag{5.10}$$

The example has illustrated the calculation of an effect size in a specific situation. Meta-analysis is concerned with combining such estimates from a number of studies and explaining differences between those estimates. The next paragraph is devoted to one of the earliest meta-analytical technique: vote-counting. This technique was developed as an alternative to usual literature surveys.

5.3 Vote-counting

Literature reviewers often employ vote-counting procedures to assess whether a specific effect does or does not exist. When reviewing a body of studies which make study of a similar phenomenon they often count the number of studies with a significant effect size. They often use this to conclude that the effect under study was significant if a majority of the studies (or a prespecified percentage of the studies) reported a significant result. The vote-counting procedure examined here is based on the same idea, but is statistically more rigorous.

The statistically advanced vote-counting procedure has a major advantage: to apply it requires little information from the research reports. Even if not every research report mentions the estimate of the effect size, it is usually possible to derive an assumed common effect size (notated by θ) and a confidence interval for this effect (Hedges and Olkin, 1985, Cooper and Hedges, 1994). To apply the method of vote-counting it is sufficient that the research reports mention that the effect size θ_i, i=1, ..., k significantly differs from zero at a certain significande level.

Suppose that there are k independent studies which estimate the same effect θ. Independency of studies implies that the different studies are concerned with different groups of people or, in other words, with different populations. These estimates are successively t_1, ..., t_k. All research reports mention whether their estimates are significant or are not significant at a certain significance level α; this is the case when t_i is larger than the critical value $C_{\alpha i}$.

It will be assumed that the probability distributions of the estimators T_i of θ_i, i=1, ..., k, only differ from each other because the samples in the individual studies have a different size.

If all sample sizes are equal, the method of "vote-counting" can be carried out in two steps. The first step concerns the actual counting of the votes. This step estimates the parameter p, the probability of a significant result (i.e. one vote for the existence of the effect θ). Here p is interpreted as the percentage of significant results.

Consider the studies as k independent trials. A trial is a success if $t_i > C_{\alpha i}$; otherwise the trial is a failure. Because the sample size of each study is the same, and consequently the estimators T_i are equally distributed, the critical value C_α is the same for all studies.
Now define the variable X_i as:

$X_i=1$, if trial i is a success, i.e. $t_i>C_\alpha$;
$X_i=0$, if trial i is a failure, i.e. $t_i \leq C_\alpha$.

The variable X_i has the following probability distribution:

$P(X_i=1) = P(T_i>C_\alpha) = p(\theta) = p$, for i=1, ..., k;
$P(X_i=0) = P(T_i \leq C_\alpha) = 1-p(\theta) = 1-p$, for i=1, ..., k.

X_i has an alternative distribution with parameter p; ΣX_i has consequently a binomial distribution with parameters k and p. The maximum likelihood estimator of p is

$$p^* = \sum_{i=1}^{k} \frac{X_i}{k} \tag{5.11}$$

With the help of the estimator p^* a confidence interval of p can be constructed. An exact interval is obtained by using the binomial distribution of ΣX_i. It is usually easier to approximate the interval using the fact that, if $kp \geq 5$ and $k(1-p) \geq 5$, the estimator p^* is approximately normally distributed with mean p and variance $p(1-p)/k$.

The second step converts the estimate of p into an estimate of the common effect size θ. The estimator of p is a function of the estimators of θ, T_i:

$$p^* = \frac{\sum_{i=1}^{k} X_i}{k} = \frac{\sum_{i=1}^{k} I_i(T_i)}{k}, \tag{5.12}$$

in which $I_i(T_i)$ is the indicator function that is defined as:

$I_i = 0$, if $T_i \leq C_\alpha$;
$I_i = 1$, if $T_i > C_\alpha$.

There are tables available in which, if one knows the values of p^* and n, one can find the corresponding value of the estimate of θ. There are tables for the following measures of effect: the standardized mean difference (Hedges and Olkin, 1985) and the sample correlation (Hedges and Olkin, 1985). Also using these tables, the boundaries of the confidence interval of p can be converted into the boundaries of the confidence interval θ.

The two-step-method discussed above can only be applied if the sample sizes in all studies are all equal. If the sample sizes are not equal, the estimators T_i of θ do not have the same probability distribution; the critical value is different in each study:

$P(X_i=1) = P(T_i > C_{\alpha i}) = p_i(\theta,n_i) = p_i$, for i=1, ..., k;
$P(X_i=0) = P(T_i \leq C_{\alpha i}) = 1\text{-}p_i(\theta,n_i) = p_i$, for i=1, ..., k.

The X_i's are independent but they do not have the same probability distribution. Consequently, ΣX_i does not have a binomial distribution any longer.

If the sample sizes are unequal but do not differ very much, the two-step-method can still be applied, as if the sample sizes were equal. Then a specific average of the sample sizes is chosen as a common sample size. The following average is unsensitive to large sample sizes:

$$\bar{n} = \left(\frac{\sqrt{n_1} + \ldots + \sqrt{n_k}}{k} \right)^2 \tag{5.13}$$

If the samples have very different sizes, however, θ can be estimated by its maximum likelihood estimator. The log-likelihood is:

$$L(\theta) = X_1 \log(p_1(\theta)) + (1-X_1)\log(1-p_1(\theta)) + \ldots + X_k \log(p_k(\theta)) + (1-X_k)\log(1-p_k(\theta)) \tag{5.14}$$

There is not usually an analytical expression for the estimator of θ, so that the maximum likelihood estimate θ is calculated numerically.

In conclusion, the vote-counting procedure discussed is statistically more advanced than the conventional vote-counting-procedure. The latter only counts the number of significant results. In case of equal sample sizes, the vote-counting procedure gives a confidence interval for the percentage of significant results. In addition, it is possible to calculate an estimate of the assumed common effect size, even if the research reports do not provide estimates of the individual effect sizes.

5.4 Combined Significance

The test of combined significance meets the wish of many literature reviewers to reach a conclusion concerning the existence of the effect under study. The hypothesis tested states that the effect is absent in all studies considered. Rejection of this hypothesis supports the existence of the effect.

Suppose that there are k independent studies. Study i tests the null-hypothesis (H_{0i}) that a specific effect θ_i is 0. The test statistic is T_i. The probability that T_i, given H_{0i}, is larger than the value t_{0i} which has been observed during the experiment, is called the (right-)p-value (p_i):

$$p_i = P(T_i \geq t_{0i}) \tag{5.15}$$

A small value of the p-value means that the observed value of the test statistic is unlikely when the null-hypothesis is true. A p-value smaller than a prespecified significance level α, leads to rejecting the null-hypothesis. Then, θ_i differs significantly from 0, or in other words, θ_i is significant.

If the test statistic T_i is continuous - which is true if the test statistic is a function of continuous random variables -, the p-value p_i has a uniform [0,1] distribution under the null-hypothesis. Consequently, the probability distribution of a p-value is not dependent on the parametric probability distribution of the observations.

The test of combined significance tests the null-hypothesis that the effect is 0 in all studies i=1, ..., k. In other words, it tests the null-hypothesis that the effects or phenomena of interest are not present in any of the populations studied; this means: H_0: $\theta_1=...=\theta_k=0$ (Cooper and Hedges, 1994; Hedges and Olkin, 1985; Wolf, 1986).

The test of combined significance is based on the p-values p_i. Because the p_i's have all the same probability distribution, the test can be applied to a collection of studies which use different models and different test statistics.

The area of the alternative hypothesis of the test of combined significance is very large: H_1: $\theta_1\neq0$, ..., $\theta_k\neq0$.

Rejecting H_0 does not say anything about the size and the sign of the effect.

A test which is best for all alternatives of H_0, does not exist; there is not a uniformly most powerful test. When a researcher chooses a test, he must take account of the deviations with respect to H_0 which he expects to occur. He must choose the test which is most powerful for the parameter values in the alternative area that he considered as most likely. In this respect, Appendix 5.1 pays attention to some criteria of optimality for tests of combined significance.

The collection of tests of combined significance can be divided into two groups. The first group consists of methods which are directly based on the uniform distribution of the p-values. An example is Trippet's method from 1931, which is also called the minimum p method. The test statistic of this test is the smallest of the k p-values: $p_{[1]}$. If the significance level is equal to α, the critical value is equal to

$$C_\alpha = 1-(1-\alpha)^{\frac{1}{k}} \tag{5.16}$$

H_0 will be rejected, if $p_{[1]} \leq C_\alpha$.

The minimum p method is a special case of a more general method: Wilkinson's method from 1951. This test does not use the smallest of the k p-values as the test statistic, but it uses the r-th order statistic: $p_{[r]}$. This $p_{[r]}$ has a beta distribution with parameters r and k-r+1.

The second group consists of methods which transform each p-value into a value of a random variable whose distribution is different from the uniform distribution. This transformation is carried out by an inverse distribution function $F^{-1}(p)$. The inverse χ^2 method (also called Fischer's method from 1932) takes the product of all k p-values as starting-point. It transforms this product, x, using the function -log(x). Next it makes use of the proposition that -2log(U) has a χ^2-distribution with 2 degrees of freedom if U is uniformly (0,1) distributed. H_0 will be rejected, if

$$-2\sum_{i=1}^{k} \log(p_i) \geq \chi^2_{2k;\alpha} \tag{5.17}$$

$\chi^2_{2k;\alpha}$ is the critical value of the χ^2-distribution with 2k degrees of freedom at a significance level of α.

Another test belonging to the second group is the inverse normal-method which is also called Stouffer's method from 1949. This method transforms each p-value into a value Z_i which comes from a standard normal distribution: $Z_i = \Phi^{-1}(p_i)$. Next it adds up the individual Z_i's; this sum has a normal (0,k) distribution. The test uses the test statistic

$$Z = \frac{Z_1 + \ldots + Z_k}{\sqrt{k}}, \tag{5.18}$$

which has a standard normal distribution. H_0 will be rejected for large values of Z.

The test of combined significance is a relatively little informative meta-analytical tool. It only provides insight in the existence of the effect while it does not provide any insight into the size and the sign of the effect. The test is particularly useful for combining outcomes of studies which examine a similar kind of effect but use different statistical models for this purpose. It should again be noted that the test of combined significance is a non-parametric method (Hedges and Olkin, 1985).

5.5 Combining Effect Sizes and the Test of Homogeneity

Suppose that there are several studies available which examine the size of the effect of the same treatment or therapy. The studies considered by the meta-analyst are assumed to be independent, implying that they are concerned with different populations. The research reports provide estimates of the effect size as well as measures of its accuracy: the estimated variance of the effect size.

Each study provides information about the same effect. Differences between the estimates can be partially attributed to the fact that the studies use different samples of the total population and partially to other reasons like the different conditions under which the research took place. Attention is paid here to the question of how different estimates can be combined into one estimate of the effect size. Consideration will also be given to whether it is justified to do this. The test of homogeneity can help judge this. It tests the hypothesis that in every study the same real effect size is examined. If this hypothesis is rejected, one should not reason from the fixed effects model without taking account of factors that explain the differences between outcomes. Section 5.6 will deal with explanatory factors of these differences.

The fixed effects model assumes that there exists one common effect size in all done examinations. In this case combining the found effect sizes means computing an estimate of the size of this single common effect. If the random effects model is assumed to be the correct model, there does not exist one single common effect size in all studies. The size of the real effect is different in each study. In the case of random effects, combining found effect sizes means assessing the average size of the real effect.

Studies differ in many respects, for instance in the size of the sample and the accuracy of the observations. As a consequence, the results of some study will be more reliable than the results of others. A meta-analyst combining individual effect sizes, should take account of this.

A good estimate of the common effect in case of the fixed effects model, or the average effect size in case of the random effects model, will be a weighted average of the found effect sizes. The estimates found in high quality studies should be given a larger weight than estimates found in low quality studies.

Let T^c be the estimator of the common effect size and w_i the weight attached to study i. T^c is by definition equal to;

$$T^c = \frac{\sum_{i=1}^{k} w_i T_i}{\sum_{i=1}^{k} w_i} \tag{5.19}$$

The combination of weights (w1, ..., wk) is called the weighting scheme. Weighting the effect sizes aims at obtaining the most accurate estimate of the common effect size. Suppose that the estimator T_i of the effect θ_i has mean θ_i and variance v_i. The estimator T^c, which is a weighted average of the individual T_i's, gives the most accurate estimate of θ, if its variance is minimized. Such values will, therefore, have to be attached to the weights w_i so that the variance of T^c is minimized. This turns out to be true, if $w_i=1/v_i$.

Example (continued)

Suppose that there are k independent studies available which have examined the effect of a specific medicine. All studies have two groups of which the first is given the medicine and the second group is not.

In case of the fixed effects model the weight w_i in equation (5.19) is equal to the inverse of the variance of the estimator of δ_i (Hedges and Olkin, 1985). If the sample sizes are large, the variance $\sigma^2(D_i)$ from proposition 1 from Appendix 5.1 can be used for this:

$$\sigma^2(D_i) = \frac{n_{Ei}+n_{Ci}}{n_{Ei}n_{Ci}} + \frac{\delta_i^2}{2(n_{Ei}+n_{Ci})} \tag{5.20}$$

There is a problem that the weight w_i depends on the unknown parameter δ_i. This can be circumvented by approximating the weight w_i with an expression which does not depend on δ_i. If the effect sizes approach 0, the expression $n_{Ei}.n_{Ci}/(n_{Ei}+n_{Ci})$ is a good approximation to the weight w_i. An alternative is to fill in the estimate d_i for δ_i. This alternative produces the following estimator of the common effect size δ:

$$D^+ = \sum_{i=1}^{k} \frac{D_i}{\sigma^2(d_i)} / \sum_{i=1}^{k} \frac{1}{\sigma^2(d_i)} \tag{5.21}$$

D^+ is a biased estimator of δ; it underestimates δ, because the weight w_i will be smaller if the found d_i is larger.

A confidence interval of the common effect size δ can be constructed with the help of the probability distribution of D^+. Proposition 3 in Appendix 5.1 gives the large sample distribution of D^+.

If the effect sizes are random, the individual effect sizes can be combined to an average effect size Δ_g (Hedges and Olkin, 1985). Then, the weight w_i in equation (5.19) is equal to the inverse of the unconditional variance of the estimator D_i:

$$v_i = \sigma^2(\Delta)+\sigma^2(D_i|\delta_i) \tag{5.22}$$

Section 5.2 explained how the variance v_i can be estimated.

If the number of studies k is large and the individual effect sizes δ_i come from a normal distribution, the weighted average with weights w_i approximately has a normal distribution with mean Δ_g and variance $[\Sigma(1/v_i)]^{-1}$.

Until now no attention has been paid to the question of when the random effects model and when the fixed effects model are appropriate. Testing the hypothesis of the fixed effects model is one way to select. It is already known that this hypothesis maintains that the real effect sizes are equal in all studies. Alternatively formulated, this hypothesis states that the variance of the real effect sizes $\sigma^2(\Theta)$ is equal to 0. Rejecting this hypothesis gives reason to use the random effects model rather than the fixed effects model.

The test of this hypothesis is called the homogeneity test (Cooper and Hedges, 1994; Hedges and Olkin, 1985):

H_0: $\theta_1=...=\theta_k=\theta$, or alternatively formulated: $\sigma^2(\Theta)=0$;
H_1: $\theta_i \neq \theta$, for at least one i, or $\sigma^2(\Theta) \neq 0$.

The test statistic is equal to

$$Q = \sum_{i=1}^{k} \frac{(T_i - T_w)^2}{v_i}. \tag{5.23}$$

If the sample size is large in each study, Q asymptotically has a χ^2-distribution with k-1 degrees of freedom under H_0. H_0 will be rejected if the value of Q is large.

Example (continued)

The question whether the size of the effect of the medicine is the same in all studies, can be answered by the test of homogeneity:

H_0: $\delta_1=...=\delta_k=\delta$;
H_1: $\delta_i \neq 0$, for at least i.

When the sample sizes are small, a test statistic for this test is unknown. However, the test statistic from equation (5.23) is accurate enough, if $n_{Ei} \geq 10$ and $n_{Ci} \geq 10$ for each study i and $1{,}5 \leq \delta \leq 1{,}5$. Then, in equation (5.23) T_i is substituted by the estimator D from the fixed effects model, T^c is substituted by the estimator D^+, and finally v_i is substituted by the conditional variance $\sigma^2(D_i | \delta_i)$ (Hedges and Olkin).

5.6 Analyzing Effect Sizes

Rejecting the hypothesis that the real effects sizes are equal in all studies, gives rise to a question which is very interesting to meta-analysts: what is the cause of the variations between the found effect sizes? Meta-analysts want to explain these variations. Estimation of a linear regression model that associates the found effect sizes with a number of independent variables, is a way to explain these variations. Those independent variables are called moderator variables.

Suppose that there exists a linear combination of the p variables, x_1, ..., x_p, which completely explains the variations in the real effect sizes (Cooper and Hedges,

1994). As a consequence, the effect size is known if these p variables are known. Then, the effect size is fixed and not random:

$$\theta_i = x_{i1}\beta_1 + x_{i2}\beta_2 + \dots + x_{ip}\beta_p \qquad (5.24)$$

This is the reason that this model is a fixed effects model, although the real effect sizes are different in each study.

The real values of the θ_i's are not known, unlike the estimates of $t_1, \dots, t_k$. Let ε_i be the error resulting from estimating θ_i:

$$\varepsilon_i = t_i - \theta_i \qquad (5.25)$$

From this and the model of θ_i, the regression equation can be derived as:

$$t_i = x_{i1}\beta_1 + x_{i2}\beta_2 + \dots + x_{ip}\beta_p + \varepsilon_i \qquad (5.26)$$

If the effect sizes θ_i have been estimated by the unbiased estimators T_i, which are normally distributed with variance v_i, the disturbances ε_i do not have equal variances. In other words, the observations $t_1, \dots, t_k$ are heteroscedastic. As a consequence, the ordinary least squares estimators of the parameters $\beta_1, \dots, \beta_p$ are unbiased and consistent, but they are not asymptotically efficient. However, the generalized least squares estimators of $\beta_1, \dots, \beta_p$ are unbiased, consistent and also asymptotically efficient.

Example (continued)

Replacing t_i by the estimate d_i in model (5.26) leads to (Hedges and Olkin, 1985):

$$d_i = x_{i1}\beta_1 + x_{i2}\beta_2 + \dots + x_{ip}\beta_p + \varepsilon_i \qquad (5.27)$$

The variance-covariance matrix of the disturbances ε_i is approximated by the diagonal matrix $\Sigma_d = (\sigma^2(d_1), \dots, \sigma^2(d_k))$, with the variance $\sigma^2(d)$ as defined in proposition 1 (see Appendix 5.1). The exact variances v_i are unknown.
The generalized least squares estimator of β,

$$\beta^* = (X'\Sigma_d^{-1}X)^{-1}X'\Sigma_d^{-1}d,$$

is now asymptotically normally distributed with mean β and a variance-covariance matrix $(X'\Sigma_d^{1}X)^{-1}$.

Rarely does a set of p variables exist of which a linear combinations can explain the variations in the found effect sizes perfectly (Cooper and Hedges, 1994). In most cases, therefore, a model that takes account of the imperfections of the explanatory model is more realistic:
where u_i is the difference between the predicted effect size and the real effect size,

$$\theta_i = x_{i1}\beta_1 + x_{i2}\beta_2 + \ldots + x_{ip}\beta_p + u_i \qquad (5.28)$$

and may be considered as a specification error. The u_i's, i=1, ..., k, are equally distributed with mean 0 and variance σ_θ^2. Model (5.28) is a mixed effects model with fixed effects β_1, ..., β_p and the random effect u_i (θ_i). The regression equation, which not only accounts of the measuring error ε but also takes account of the imperfections of the model (u_i), is:

$$t_i = x_{i1}\beta_1 + \ldots \beta_p x_{ip} + u_i + \varepsilon_i \qquad (5.29)$$

The variance of the disturbance $u_i+\varepsilon_i$ is equal to the estimated variance v_i of the effect size plus the unknown variance σ_θ^2:

$$\mathrm{var}\,(u_i + \varepsilon_i) = v_i + \sigma_\theta^2 \qquad (5.30)$$

v_i is mentioned by the research reports, but σ_θ^2 is unknown and has to be estimated. The estimators of the parameters β_1, ..., β_p depend on the unknown variance σ_θ^2. Model (5.29) is estimated using either the method of moments or the maximum likelihood method (Cooper and Hedges, 1994).

5.7 Conclusions

In conclusion, the meta-analytical methods generally used by medical and social scientists are classical statistical methods. Ordinary statistical analysis and meta-analysis differ with regard to the kind of data which they describe and interpret. Ordinary statisticians deal with original observations, while meta-analysts deal with data which were summarized by other people before. These summarized data usually represent effect sizes. These are effects of, for instance, a specific medicine or psychotherapy.

To apply the meta-analytical statistical methods discussed above it is necessary that the individual studies make use of statistical models. Observations which are gathered by the individual researchers, should be assumed to come from a probability distribution, so that the reported estimate of the effect size comes from a probability distribution itself and the accuracy of this estimate can be expressed by a confidence interval or variance. Studies which do not provide estimates of the accuracy of effect sizes in this form are less suitable for the application of the methods discussed here, because combining estimates and carrying out weighted regression analysis is based on this type of information.

Appendix 5.1 Background Propositions

In this annex we will concisely give some statistical details on Sections 6.2, 6.4 and 6.5.

Proposition 1 (Hedges and Olkin, 1985):
If the ratios n_E/N and n_C/N remain constant when $N=n_E+n_C$ increases, D has asymptotically a normal distribution with mean δ and variance $\sigma^2(\delta)$, with:

$$\sigma^2(\delta) = \frac{n_E+n_C}{n_E n_C} + \frac{\delta^2}{2(n_E+n_C)} \qquad (5A.1)$$

$\sigma^2(\delta)$ is estimated by substituting the estimate d for δ. The estimate d is obtained with the help of D.

As mentioned in Section 5.4, there does not exist one best test of combined significance. It is possible to compare tests of combined significance by means of some criteria of optimality. These criteria refer to the behaviour of the test statistic. However, they do not give a complete picture of the performance of any statistical test. Two important criteria of optimality of which you should take notice while choosing a test of combined significance are (Hedges and Olkin, 1985):
1) admissibility,
2) Bahadur-optimality.

Definition 1:
A test of combined significance is admissible, if it is the most powerful test for some alternative H_1.

Definition 2:
A test of combined significance is monotone, if, whenever the set of significance levels $(p_1, ..., p_k)$ leads to rejection of H_0, every set of significance levels $(p_1^*, ..., p_k^*)$, such that $p_1^* \leq p_1, ..., p_k^* \leq p_k$, leads to rejection of H_0.

Result:
A monotone test is admissible.

Proposition 2:
If the distribution of each of the test statistics $T_1, ..., T_k$ is the 1-parameter exponential family, a test of combined significance is admissible if and only if the acceptance region is convex.

Bahadur-optimality refers to the large sample behaviour of the combined significance test statistic.

Definition 3:
A test is Bahadur-optimal, if it gives the most significant results when the sample size is large. That is, the test will have the smallest possible significance level for large samples. The following table shows which tests described in Section 5.4 are admissible or Bahadur-optimal.

	admissible for 1-parameter-exponential family	Bahadur-optimal
minimum-p method	yes	no
Wilkinson's method	no	no
inverse-χ^2 method	yes	yes
inverse-normal method	yes	no

Proposition 3 (Hedges and Olkin, 1985):
If the ratios n_{Ei}/N and n_{Ci}/N, i=1, ..., k, remain constant when $N=\Sigma(n_{Ei}+n_{Ci})$ increases, the estimator D^+ is asymptotically normally distributed with mean

$$\delta^+ = \sum_{i=1}^{k} \frac{\delta_i}{\sigma^2(d_i)} / \sum_{i=1}^{k} \frac{1}{\sigma^2(d_i)}, \tag{5A.2}$$

and variance

$$\sigma^2(d^+) = \left(\sum_{i=1}^{k} \frac{1}{\sigma^2(d_i)} \right)^{-1} \tag{5A.3}$$

CHAPTER 6 MEASUREMENT AND UNCERTAINTY ISSUES IN ENVIRONMENTAL ECONOMICS AND DECISION ANALYSIS

6.1 Introduction

This chapter deals with the relationship between uncertainty, measurement theory, and decisions in the context of meta-analysis. The potential usefulness of meta-analysis in environmental science and economics may be particularly relevant for research cases which incorporate imprecision in information.

Economics deals with issues which often involve some kind of risk or uncertainty. Although there are notable exceptions and developments which specifically focus on incorporating risk and uncertainty elements, deterministic equilibrium approaches have dominated economic science in this century. Such approaches do not regard economic decisions as being made under conditions of pervasive uncertainty and incomplete knowledge. Once it is decided that it is important, uncertainty should be incorporated in an analysis. This does not necessarily mean, though, that it is specified in terms of stochastic models of the type as developed in the fields of decision analysis and operational research. An alternative approach could adopt a disequilibrium perspective where there is a continuous gap between expectations and realizations. Uncertainty can, however, also be dealt with on a more fundamental level. When it is taken for granted that economic decisions are always made under differing psychological conditions, then one may assume a fundamental uncertainty and unpredictability in the decisions made by economic agents. In any case, statistical inference from different studies undertaken under varying conditions provokes many questions on the validity of comparative analysis or on the transferability of results.

This chapter makes issues of uncertainty in economic decision models more transparant by describing theoretical foundations of the properties of vague data sets. It focuses on two advances in research on this subject, comprising research on stochastic elements and research on fuzzy data sets. Such approaches are particularly relevant in the fields of environmental economics and environmental science.

6.2 Uncertainty in Economics and Decision Theory

6.2.1 Types of uncertainty

Uncertainty is a generic term which embraces both risk and strict uncertainty. Economists have traditionally focused on risk where there is a known probability distribution and which can be treated within normal stochastic revealed preference frameworks. Strict uncertainty poses more problems because there are no such known distributions. There have been increasing efforts in economics to handle it through stated preference models. Central in the approach to risk is the concept of contingent consumption claims: goods are defined to have, in addition to the normal physical and service attributes, also a subscription to states of nature. For such goods expected utility theory, based on maximizing expected utility, was developed (von Neumann and Morgenstern, 1944). This approach is directly based on the rationality postulate which is central to neoclassical economic analysis. This postulate is defined by Blaug (1992) as: "choosing in accordance with a preference ordering that is complete and transitive, subject to perfect and costlessly acquired information".

Two forms of uncertainty are normally dealt with, *viz.* market uncertainty and technological uncertainty (Hirshleifer and Riley, 1979). In the first concept individuals are uncertain about the actions of other economic agents. At the micro-economic level, search processes dominate, while at a macro or meso level market disequilibrium and price dynamics are essential. This form of uncertainty concerns the endogenous variables of the economic system and is institutionally induced. Event uncertainty, on the other hand, deals with exogenous data, *viz.* resource endowments and production possibilities. In the case of exhaustible resources, it is linked with such things as size and quality of resource reserves, costs of extraction, discovery of new reserves and invention of substitute products.

Decision-makers can take either terminal actions or informational actions. The first actions will allow individuals to adapt to uncertainty, while the second points individuals to overcome uncertainty. Within the class of terminal actions, we can distinguish between the trade, sharing or modification of risk. Informational decision-making starts with the acquisition of an information service followed by Bayesian updating of subjective prior probabilities assigned to states of the world (DeGroot, 1970). In addition, information may also emerge autonomously with the mere passage of time. Then a trade-off between costs of waiting and irreversibility of some development is possible.

Values of informational activities are based upon the expected utility gains from shifting to better choices among the set of terminal actions. A specific informational activity relevant to long run issues such as irreversibility and environmental sustainability is an indirect one; in other words waiting, in order to be able to enjoy the option value (Usategui, 1990; Tu and Wilman, 1992). This option value is defined as the gain from being able to learn about future benefits, that would be precluded by some development, if one does not initially develop, but decide in favour of preservation. Therefore, an option value is a conditional value.

For incorporating risk in an analysis, there is a large body of techniques, models and rules, ranging from sensitivity analysis, probability analysis or decision-theoretic modelling. Rules of thumb may be used as well, such as adapting discount rates for uncertainty, limiting the time horizon in dynamic analysis, or restricting the space dimensions of the problem under consideration. Searching for surprises, for instance in the context of complex systems or long run problems, may be based on forum or expert techniques, such as Delphi methods.

In order to arrive at mathematically sound results in dealing with the behaviour of complex systems in time, various model types can be used (Bennet and Chorley, 1978). In the context of optimal modelling, stochastic dynamic optimal control techniques are suitable when a change in values of stock variables over time can be described by a process which has a drift and a diffusion component. The major shortcoming in such advanced techniques is that they can handle only a very limited amount of complexity and multidimensionality. Stochastic simulation with a nonlinear model and re-estimation of a linear model while iteratively finding better controls is an approach to incorporating uncertainty. In these cases the parameters may change over time so as to allow for making small and drastic changes reflected in in the model structure. The advantage of feedback control structures, necessary for control of stochastic systems, is the fact that the system behaviour under a given control structure can be studied easily, for both deterministic and stochastic systems (Chow, 1975 and Segerson, 1988).

Alternatively, in deterministic optimal control the impact of disturbances in the paths of state variables on the value of the objective function can be investigated by means of simulation modelling. It makes a considerable difference whether or not the optimal control path is very sensitive to slight disturbances in the paths of state variables. It is also important to know the character of stationary points and the sensitivity of equilibria. If more equilibria for the dynamic behaviour of the system output exist, it is necessary to know how the system can move from one to another equilibrium and what this implies for the value of the objective function. Also important is the study of the combined effect upon the system's behaviour and performance when multiple states behave in an uncertain way. Sensitivity analysis with respect to initial and end conditions on state variables, parameter values in objectives, state transition equations and constraints are meaningful when uncertainty on their specific values exists. It is noteworthy that in many cases optimal control models are presented as if a fixed structure is taken for granted. This may be at odds with long-term changes in parameter structures, so that in that case a blend between optimal control theory and chaos theory may be useful (Kelsey, 1988).

6.2.2 Types of uncertainty in environmental and resource economics

In economics dealing with environmental policy issues, the issue of uncertainty may even be more pervasive than in standard microeconomic topics (Ploudre and Yeung, 1989). This may be linked to two central features: an interest for long term issues often seen in terms of sustainability, and identification, analysis and control of complex systems (environmental systems on various spatial levels). New conceptual approaches try also to deal with uncertainty, such as hierarchical systems and network arrangement of subsystems. In the view of Walters (1986) resource managers must learn to live with substantial uncertainties and are advised even to experiment with reality in order to learn and be able to improve resource management practices. Modelling can clarify these, by decomposition of relationships into smaller and more understandable pieces. In the end, management still comes down to gambling. This does not sound attractive, but offers in any case opportunities for learning games.

Finally, in the context of evaluation, uncertainty is often discussed with regard to choice of the discount rate in benefit-cost analysis. Temporal aspects of uncertainty are then stressed. In the context of multicriteria evaluation uncertainty has also recently received some attention (Munda, 1995). This includes a broader discussion of system uncertainty.

By way of summarizing, the types of uncertainty in the context of economic-environmental-institutional system analysis may be categorized as follows (Bennet and Chorley, 1978; Walters, 1986; Funtowicz and Ravetz, 1990):

On the fundamental level:
- Randomness relating to the limits of causality and determinism.
- Limits of correspondence between descriptive categories and the reality to which they refer.

On the level of modelling:
- Which variables to focus on: indicators, exogenous and endogenous variables? This uncertainty applies to every formal approach.
- Statistical or parametric uncertainty about the form and parameter values of various functional responses. In statistical approaches one may use procedures for finding one's way in the space of possible functional forms, but this requires high quality and large quantity data. Since usually for socio-economic and environmental systems controlled experimentation is very limited or even impossible, this approach will be rather imperfect.
- Errors in terms of exactness of measurement, vagueness and fuzziness in information.

On the level of implementation:
- Unpredictability and uncontrollability of economic and environmental processes. These force to use feedback policies, based on monitoring and adjustment to changes, or to wait and learn. In addition, there may be delays between analysis and implementation.

In order to effectively develop and analyse policy strategies it should be clear how systems work. This involves fundamental issues about cause-effect relationships and behavioural mechanisms. The causal structure of environmental systems provides many difficulties for analysts, because of lack of information, knowledge, incomplete data or measurement errors. For this reason one may focus on uncertainty analysis, i.e. making use of deterministic dynamic simulation with sensitivity analysis, stochastic simulation, or Monte Carlo simulation (Hettelingh, 1989).

6.3 Measurement Theory

The process of grouping individual observations into qualitative classes is measurement at its most primitive level. Sometimes this is called categorical or nominal scaling. The set of equivalence classes itself is called a nominal scale. The word measurement is usually reserved for the situation in which each observation is assigned a number, with this number reflecting a magnitude of some quantitative property. The question how to assign this number constitutes the so-called representation problem.

There are at least three kinds of numerical measurement that can be distinguished: these are ordinal, interval and ratio scales (Winkler and Hays, 1975). Imagine a set of objects O, and that there is some property that all objects in the set possess such as value, length, or intelligence. Furthermore, let us suppose that each object o has a certain amount or degree of that property. In principle it is possible to assign a number $t(o)$, to any object $o \in O$, standing for the amount that o actually "has" of the characteristic at hand. Ideally, to measure an object o, we would like to determine this number $t(o)$ directly. However this is not always possible, and therefore there is a need to find a procedure for pairing each object with another number, $m(o)$, that can be called its numerical measurement. The measurement procedure used constitutes a function rule $m : O \rightarrow R$, telling how to give an object o its $m(o)$ value in a systematic way. Measurement operations or procedures differ in the information that the numerical measurements themselves provide about the true magnitudes.

Let us suppose that there is a measurement procedure or rule for assigning a number $m(o)$ to each object $o \in O$, and suppose that the following statements are true for any pair of objects o_1 and $o_2 \in O$:

$$m(o_1) \neq m(o_2) \quad \text{only if } t(o_1) \neq t(o_2) \tag{6.1}$$
$$m(o_1) > m(o_2) \quad \text{only if } t(o_1) > t(o_2) \tag{6.2}$$

In other words, by this rule is possible to say that if two measurements are unequal, and if one measurement is larger than another, then one magnitude exceeds another. Any measurement procedure for which both statements (6.1) and (6.2) are true is an example of ordinal scaling, or measurement at the ordinal level.

A fundamental point in measurement theory is the uniqueness of scale, i.e. what are the admissible transformations of scale allowing that the truth or falsity of the statement involving numerical scales remain unchanged (problem of meaningfulness). In the case of an ordinal scale, it is unique up to a strictly monotone increasing transformation (with infinite degrees of liberty). Other measurement procedures associate objects $o \in O$ with a real number $m(o)$ where much stronger statements can be made about the true magnitudes from the numerical measurements. Suppose that the following statement, in addition to statements (6.1) and (6.2) is true:

$$t(o) = x \text{ if and only if } m(o) = ax + b, \text{ where } a \in R^+ \tag{6.3}$$

That is, the numerical measurement $m(o)$ is some affine function of the true magnitude x. When the statements (6.1), (6.2) and (6.3) are all true, the measurement operation is called interval scaling, or measurement at the interval-scale level. An interval scale is unique up to a positive affine transformation (with two degrees of freedom).

When measurement is at the interval scale level, any of the ordinary operations of arithmetic may be applied to the differences between numerical measurements, and the results can be interpreted as statements about magnitudes of the underlying property. The important part is the interpretation of a numerical result as a quantitative statement about the property shown by the objects. This is not possible for ordinal-scale numbers, but it can be done for differences between interval-scale numbers. Interval scaling is about the best we can do in most scientific work, and even this level of measurement is all too rare in social sciences. In the physical sciences, however, it is sometimes possible to find measurement operations making the following statement true:

$$t(o) = x \text{ if and only if } m(o) = ax, \text{ where } a \in R^{+} \tag{6.4}$$

When the measurement operation defines a function such as statements (6.1) through (6.4) are all true, then measurement is said to be at the ratio-scale level. For such scales, ratios of numerical measurements are unique and can be interpreted directly as ratios of magnitudes of objects. A ratio scale is unique up to a linear transformation; in this case, the ratio between differences is unique with only one degree of freedom.

Of course, the less the admissible transformations of a scale, the more meaningful are the statements involving that scale. From this point of view, it is better to have a ratio scale than an interval scale, and it is better to have an interval scale than an ordinal scale. In the following sections for simplicity, quantitative information is referred to when the measurements are on a interval or ratio scale, and to qualitative information when the measurements are on a nominal or ordinal scale.

6.4 Fuzzy Uncertainty in Decision Models

6.4.1 Introduction to fuzzy sets

Ideally, the information available for a decision model should be precise, certain, exhaustive and unequivocal. In reality though, it is often necessary to use information which does not have these characteristics and to face the uncertainty of a stochastic and/or fuzzy nature present in the data (Munda, 1995). The combinations of the different levels of measurement with the different types of uncertainty, therefore, have to be taken into consideration (Table 6.1).

These types of information have been extensively studied in measurement theory, probability theory and statistics; but some open problems still remain, above all regarding the representation of ordinal information in evaluation and decision problems.

Fuzzy uncertainty does not concern the occurrence of an event but the event itself, in the sense that it cannot be described unambiguously. This situation is very common in human systems. Spatial-environmental systems in particular, are complex systems characterised by subjectivity, incompleteness and imprecision (e.g., ecological processes are quite uncertain and little is known about their sensitivity to stress factors such as various types of pollution). Zadeh (1965) writes: "as the complexity of a system increases, our ability to make a precise and yet significant statement about its behaviour diminishes until a threshold is reached beyond which precision and significance (or relevance) become almost mutually exclusive characteristics" (incompatibility principle). Therefore, in these situations statements as "the quality of the environment is good", "the unemployment rate is low" are quite common. Fuzzy set theory is a mathematical theory for modelling situations, in which traditional modelling languages which are dichotomous in character and unambiguous in their description cannot be used. Human judgements, especially in linguistic form, appear to be plausible and natural representations of cognitive observations. We can explain this phenomenon by cognitive distance. A linguistic representation of an observation may require a less complicated transformation than a numerical representation, and therefore less distortion may be introduced in the former than in the latter. Fuzzy sets as formulated by Zadeh are based on the simple idea of introducing a degree of membership of an element with respect to some sets. The physical meaning is that a gradual instead of an abrupt transition from membership to non-membership is taken into account.

Table 6.1. Possible combinations of information measurement levels with respect to different types of uncertainty: some exemplary topics.

	Quantitative information	*Qualitative information*
Deterministic		
Data	Ratio scales Interval scales	Ordinal scales Nominal scales
Models	Classical economic models (Neo-classical models, Keynesian models, etc.)	Case studies Non-formal descriptions Cognitive mapping
Stochastic		
Data	Probability distribution	Linguistic probabilistic evaluations
Models	Statistical models	Machine learning Brain storming Delphi model
Fuzzy		
Data	Fuzzy number Fuzzy set	Linguistic evaluation of considered variables
Models	Fuzzy extension of classical models Fuzzy logic	Machine learning Brain storming Delphi model
Soft		
Data	Multi-attribute quantitative description	Multi-attribute qualitative description
Models	Codification by means of thresholds Evidence theory	Rough set, Interval set Rule induction Modal logic Three value logic Four value logic

Fuzzy nominal information is the base of the whole fuzzy set theory, since it considers all cases between 0 (non membership) and 1 (complete membership); it is represented by means of the membership functions.

Let us assume that the symbol U represents the entire set (Universe of discourse). In classical set theory, given a subset A of U, each element $x \in U$ satisfies the condition: either x belongs to A, or x does not belong to A. The subset A is represented by a function $f_A: U \rightarrow \{0, 1\}$:

$$f_A(x) = \begin{cases} 1 & \text{if } x \in A \\ 0 & \text{if } x \notin A \end{cases} \tag{6.5}$$

The function f_A is called a characteristic function of the set A. Fuzzy sets are then introduced by generalising the characteristic function f_A. Let U again be a universe of discourse. Let $x \in U$. Then a fuzzy set A in U is a set of ordered pairs:

$$\{[x, \mu_A(x)]\}, \ \forall x \in U \tag{6.6}$$

where $\mu_A: U \rightarrow M$ is a membership function which maps $x \in U$ into $\mu_A(x)$ in a totally ordered set M (called the membership set) and $\mu_A(x)$ indicates the grade of membership of x in A. Generally, the membership set is restricted to the closed interval [0, 1]. A fuzzy set is completely determined by its membership function. For $0 < \mu_A(x) < 1$, x belongs to A only to a certain degree; thus there is ambiguity in determining whether or not x belongs to A. The physical meaning is that a gradual instead of an abrupt transition from membership to non-membership is taken into account. A classical example is that of age. Let U be the set of all non-negative integers. Let us take into consideration the primary terms young and old. These terms can be considered the label of two fuzzy sets A and B. No doubt the ages 6 or 10 are young, whereas the ages 30 or 40 are less young. Thus it is possible to define a membership function $\mu_{A_{young}}$ showing the degree of compatibility of the age x to the concept of young.

6.4.2 Linguistic variables

In traditional mathematics, variables are assumed to be precise, but when we are dealing with our daily language, imprecision usually prevails. Intrinsically, daily languages cannot be precisely characterised on either the syntactic or semantic level. Therefore, a word in our daily languages can technically be regarded as a fuzzy set. Formally, a linguistic variable is represented by a quintuple (X, T(x), U, G, M) (Leung, 1988; Zimmermann, 1986) where: X is the name of the variable, e.g. age; T(x) is the term set of X, finite or infinite, such as young, very young and so on, in an universe of discourse U. A primary term in T(x) is a term whose meaning must be defined a priori, and which serves as a basis for the computation of the meaning of the non primary terms in T(x). G is a syntactic rule by which the non-primary terms in the term set are generated. It is possible to use a context free grammar or a regular grammar; in G it is possible to find primary terms, hedges (not, very, more or less), relations (younger than, older than), conjunctions (and), and disjunctions (or).Thus computer implementation of such an approach presents a high degree of complexity and generally requires artificial intelligence oriented languages. M is a semantic rule which associates each term with its meaning (a fuzzy subset in U). Through M, a compatibility (membership) function μ: U $\rightarrow$ [0, 1] is constructed (e.g. μ_{young} shows the degree to which a numerical age is compatible with the concept of young and equivalently μ_{young} may be viewed as the membership function of the fuzzy set young).

A linguistic variable is, therefore, a fuzzy variable whose values are fuzzy subsets in a universe of discourse. The base variable of the linguistic variable is a precise variable which takes an individual value in its domain, i.e. the universe of discourse U. The domain of the linguistic variable is the collection of all possible linguistic values, fuzzy sets defined in the same universe of discourse through the base variable. In some cases, however, the fuzzy set which is assigned to the fuzzy restriction may not have a numerically-valued base variable. In order to allow a formal analysis, a mathematical translation of such linguistic propositions is needed. This can be done by means of possibility theory (Dubois and Prade, 1980).

If R(x) is a fuzzy restriction (a fuzzy restriction is a fuzzy relation which acts as a flexible constraint on the values that may be assigned to a variable), then the effect of F (a linguistic value) on X (base variable) can be expressed as

$$R(x) = F, \tag{6.7}$$

where X is a variable in U, F is a fuzzy set in U and R(x) is a fuzzy restriction imposed by F. Therefore, this fuzzy restriction may be expressed in a linguistic proposition:

$$P : X \text{ is } F \tag{6.8}$$

which generates a possibility distribution

$$\pi_X = F \tag{6.9}$$

Associated with the possibility distribution is a possibility distribution function such that

$$\text{Poss}\ (X=a) = \pi_x\ (a) = \mu_F\ (a) \tag{6.10}$$

We can, therefore, conclude that fuzzy restrictions, possibility distributions and fuzzy sets are closely related. In the qualitative information available for an evaluation or decision model, two different types of linguistic variables may be present (Munda, 1995):

(1) The meaning can be translated in a measure on an interval or ratio scale (quantitative base variable), e.g. age and distance;
(2) There is no meaning on an interval or ratio scale, and therefore the base variable is also qualitative in nature, e.g. appearance, comfort and beauty.

Type 1. If linguistic variables whose meaning can be translated in a measure on an interval or ratio scale are present in a decision model, generally it is because of a lack of information or of the right instrument of measurement. Therefore, we have a qualitative evaluation of a variable that in theory could be measured on an interval or ratio scale. So it is reasonable to suppose that it is possible to transform the qualitative information into a quantitative one with a certain degree of precision. The parameters, necessary in this case, may be easily established, because this is a case of the so-called informational fuzziness depending mostly on the subjective culture of the person in charge of the evaluation. For example, the proposition "that man is tall" may have different meanings for different people, but everybody can easily indicate the tolerance interval of his own evaluation. Such a representation takes into account some "labels" of the term set, e.g. young, young and/or very young, etc., and the problem is to find which values of the base variable are compatible with these terms. Formally, in general we know that

$$\text{Poss}\ (X{=}a) = \pi_x\ (a) = \mu_F\ (a) \tag{6.11}$$

Thus, if for example, the linguistic proposition is "the distance is long", then

$$\text{Poss}\ (\text{distance}{=}10\ \text{km}) = \pi_{\text{distance}}\ (10) = \mu_{\text{long}}\ (10) = 0.6 \tag{6.12}$$

indicates the possibility of 10 km being considered as a long distance. That is, the possibility distribution indicates whether or not the proposition "the distance is long" is possible, and equivalently it denotes the degree of compatibility of 10 km to the fuzzy set long. The problem may, therefore, be easily solved by establishing possibility distributions, e.g. π_{distance} = long (Leung, 1988),

$$\pi_{\text{distance}}(x) = \mu_{\text{long}}(x) = \left\{ \begin{array}{ll} 0 & \text{if } x \leq \beta \\ 1-e^{-k\left(\frac{x-\beta}{\beta}\right)^2} & \text{if } x > \beta\ (k>0) \end{array} \right\} \tag{6.13}$$

Type 2. In the case of linguistic variables with no meaning on an interval or ratio scale, the qualitative information does not depend on lack of information, but on the nature of the information that is essentially fuzzy (intrinsic fuzziness). Therefore, whereas in the other cases the stochastic representation and the fuzzy one may be competitive, in this case the fuzzy representation is the only one possible. For example, if linguistic propositions clearly have no quantitative base variable, how can we represent them? It seems that there is a set of hidden and fuzzy standards in one's mind in a justification for these types of concept, but they are more than a human being can rationally handle simultaneously.

A first approach to this problem may be to decompose the concept that one wants to represent into a series of quantitative measurable variables. This approach presents two main problems, namely the explication of the quantitative variables, and the aggregation procedure to be used.

A second approach is to define an artificial quantitative base variable, assuming that the real space is one-dimensional. The interval of the real space is chosen from [-1, 1], [0, 1], or [0, 10] and can be subdivided into a series of fuzzy sets representing linguistic values (e.g. very negative, moderately positive or very positive). Then, a link or mapping between quantitative and qualitative values is established. This "direct estimation" approach has been criticised because of its lack of theoretical foundation. Recently, some psychologists (Norwich and Turksen, 1982; Wallsten *et al.*, 1986) have developed a graded pair comparison procedure, which allows simultaneous testing of the necessary axioms and scaling of the responses in order to obtain memberships and tests of goodness of fit. Subsequently, empirical experiments have demonstrated a high level of similarity between membership values determined through graded pair comparison and direct magnitude estimation. Thus it seems that a theoretical justification for the quantification of the vague meanings of inexact linguistic terms by means of direct estimation can be established.

A third interesting approach can be the notion of type 2 fuzzy set. A type 2 fuzzy set is a fuzzy set whose membership values are fuzzy sets on [0, 1]. This corresponds to the case that the decision-maker is not able (or not willing) to characterise the grade of membership by an exact number, but gives an evaluation such as "the grade of membership is high or medium". It is always possible to define a fuzzy set of type n=2,3, ..., if its membership function is a mapping from U to a set of fuzzy subsets of type n-1. It is, therefore, possible that to reduce the fuzziness, many transformations are requested, thus diminishing drastically the computational efficiency of the algorithm.

Another possible approach has been developed in the field of psychological research (Hersh *et al.*, 1979), called the yes-no paradigm. Here, an element x of the universe of discourse U, is presented to the subject and he has to decide whether the element is a member of A, A being a fuzzy subset of U. The fraction of positive responses across replications (within or across subjects) is considered a measure of $\mu_A(x)$. The main problem of this approach is that it confounds fuzziness with response variability and it can be interpreted as an indication that words have various, but nevertheless precise meanings to different people and/or different times.

6.4.3 Fuzzy numbers

Fuzzy quantitative information has been deeply studied in the fuzzy literature by means of the notion of fuzzy numbers (Dubois and Prade, 1980). A fuzzy number is simply a fuzzy set in the real line and is completely defined by its membership function such as

$$\mu: R \rightarrow [0, 1] \qquad (6.14)$$

For computational purposes, in general this definition is restricted to those fuzzy numbers which are both normal and convex:

$$\text{normality: } \sup\{\mu(x)\} = 1 \qquad \text{with } x \in R \qquad (6.15)$$

convexity: $\mu\{\lambda x_1 + (1 - \lambda)x_2\} \geq \min\ \{\mu(x_1), \mu(x_2)\}$ (6.16)

with $x \in R$ and $\lambda \in [0, 1]$.

The requirement of convexity implies that the points of the real line with the highest membership values are clustered around a given interval or point. This fact allows one to easily understand the semantics of a fuzzy number by looking at its distribution and to associate it with a properly descriptive syntactic label (e.g. "approximately 10").

The requirement of normality implies that, among the points of the real line with the highest membership value, there exists at least one which is completely compatible with the predicate associated with the fuzzy number.

A standard normal convex trapezoidal fuzzy number can be characterised by a 4-tuple (a, b, α, δ) where [a, b] is the closed interval on which the membership function is equal to 1, α is the left-hand variation and δ is the right-hand variation. If only one point on the real line with $\mu(x)= 1$ exists, the fuzzy number is called a triangular fuzzy number.

A more general type of fuzzy number is the L-R fuzzy number; it is defined as follows:

$$\mu_A(x) = \begin{cases} F_L(x-m)/\alpha & \text{if } -\infty < x < m,\ \alpha > 0 \\ 1 & \text{if } x = m \\ F_R(x-m)/\delta & \text{if } m < x < +\infty,\ \delta > 0 \end{cases} \qquad (6.17)$$

where m, α, δ, are the middle value, the left-hand and the right-hand variation, respectively. $F_L(x)$ is a monotonically increasing membership function and $F_R(x)$, not necessarily symmetric to $F_L(x)$, is a monotonically decreasing function. If a closed interval on which the membership function is equal to 1 exists, it is called a flat fuzzy number.

6.4.4 Probability theory, possibility theory and fuzzy sets

Probabilities, possibilities and fuzzy sets are all measures used to formalise and quantify uncertainty. In particular, fuzzy sets are induced by the break down of the law of the excluded middle. This law says that if a set A is defined on a universe of discourse X, then:

$A \cup A_C = X$ and $A \cap A_C = \emptyset$ (6.18)

where A_C is the complement of A.

A direct comparison of the two theories on the grounds of their mathematical structures is quite difficult due to the fact that a unique definition of probability does not exist (e.g. classic, frequentistic, subjective and axiomatic). Therefore, if probability is defined by the concept of statistical frequency, its mathematical structure is a σ-algebra. If probability is considered to be subjective, then its mathematical structure is a Boolean ring within the system of two-value logic (Leung, 1988).

For example, let us try to compare frequentistic probability and possibility theory. The axiomatic foundation of Kolmogorov's system is a probability space (Ω, A, P). The set Ω is the outcome space and A is a system of subsets of Ω, a Borel field. The sets of A are called the events. The set Ω and the system A satisfy the following conditions:

$$\Omega \in A$$
$$\text{if } A \in \mathcal{A} \quad \text{then} \quad A_C \in \mathcal{A} \tag{6.19}$$

Let $\{A_j\}$ be a sequence of sets, if $A_j \in \mathcal{A}$, then

$$\bigcup_{j=1}^{\infty} A_j \in \mathcal{A} \text{ and } \bigcap_{j=1}^{\infty} A_j \in \varnothing, \ \forall j \tag{6.20}$$

Thus the system of subsets A is a σ-algebra. Let P(*) be a probability defined in A such as:

$$P : \mathcal{A} \rightarrow [0, 1]$$
$$P(\Omega) = 1 \tag{6.21}$$

Let $\{A_j\}$ be a sequence of events such as $A_i \cap A_j = \varnothing$, $i \neq j$, then

$$P\left(\bigcup_{j=1}^{\infty} A_j\right) = \sum_{j=1}^{\infty} P(A_j), \quad \forall j, \ A_j \in \mathcal{A} \tag{6.22}$$

Thus the set function P is σ-additive. A possibility distribution Π, is a function defined in the fuzzy power set P(X) such as:

$$\Pi : P(X) \rightarrow [0, 1]$$
$$\Pi(\phi) = 0$$
$$\text{If } A, B \in P(X), \text{ and } A \subseteq B, \text{ then } \Pi(A) \leq \Pi(B) \tag{6.23}$$

Let $\{A_j\}$ be a sequence of fuzzy sets, if $A_j \in P(X)$, then

$$\Pi\left(\bigcup_{j} A_j\right) = \sup_{j} \Pi(A_j), \quad \forall j, \ A_j \in P(X) \tag{6.24}$$

Thus, P is not σ-additive. In terms of union and intersection between events, the basic differences between probability and possibility are:

$$P(A \cup B) = P(A) + P(B), \quad \text{if } A \cap B = \varnothing, \tag{6.25}$$
$$P(A \cap B) = P(A) \circ P(B), \quad \text{if A and B are independent.} \tag{6.26}$$

$$\Pi(A \cup B) = \max [\Pi(A), \Pi(B)], \text{ if A and B are non-interactive,} \tag{6.27}$$
$$\Pi(A \cap B) = \min [\Pi(A), \Pi(B)], \text{ if A and B are non-interactive.} \tag{6.28}$$

A fuzzy system is a triple, (β, D, f), where β is a base of a Borel field that represents all the fuzzy sets of interest, D is the domain of the fuzzy sets in c(β), and f is a real valued function defined for each $A \in c(\beta)$ and each $x \in D$ such that (Drakopoulos, 1994):

$$\forall x \in D \quad f_{\varnothing}(x) = 0, \quad \varnothing \text{ is the empty set in } c(b) \tag{6.29}$$

$$\forall x \in D \quad f_{\Omega}(x) = 1, \quad \Omega \text{ is the universal set in } c(b) \tag{6.30}$$

If A_1, A_2, ... is any sequence of fuzzy sets in β then

$$\forall x \in D \quad f_{\bigcup_{i=1}^{\infty} A_i}(x) = \sup_{i=1}^{\infty} f_{A_i}(x)$$

$$\forall x \in D \quad f_{\bigcap_{i=1}^{\infty} A_i}(x) = \inf_{i=1}^{\infty} f_{A_i}(x) \tag{6.31}$$

$$\forall A \in c(\beta), \forall x \in D \quad f_{A_U}(x) = 1 - f_A(x) \tag{6.32}$$

Membership functions, seen as measures on fuzzy sets, are extensional possibility measures augmented so that they satisfy the rules defined in (6.31) and (6.32). Drakopoulos (1994) proves a number of theorems showing that probabilities are more expressive than both possibilities and fuzzy sets, though they are harder to compute. Possibilities and fuzzy sets can actually simulate probabilities but, in that case, their space requirements are exponential when compared to those of probabilities. The increased complexity of probabilities gives rise to a trade-off of complexity and capacity versus efficiency.

6.5 Conclusions

The probabilistic, possibilistic, fuzzy set and rough set approaches constitute different instruments for dealing with the various types of uncertainty encountered in the social sciences in general, and particularly in economic-environmental studies. It is exactly the different origins and characteristics of these types of uncertainty which are of importance in the choosing of the type of approach and technique best suited to deal with each. It is, therefore, essential to pay the greatest attention to the specific real world problem at hand, to its object and objectives, to the type of information available, to the level of depth necessary in the analysis, and to the type of synthesis required, so as to carry out the most suitable choice. Moreover, it is often necessary in concrete cases to accept solutions which constitute a compromise, since a trade-off must be made between the two aspects required, accuracy and efficiency.

All the instruments have their own characteristics and usefulness in dealing with the various aspects of uncertainty. Each of the approaches proves to be particularly relevant, from a technical and methodological point of view, for the analysis of a specific concrete problem. But it is exactly the different qualities of each, which demonstrates that these approaches are not intrinsically in conflict, but complementary. It is, therefore, very useful, in order to enrich and better understand the phenomenon under examination, to employ more than one method at the same time, with respect to the specific technical characteristics of each. A correct critical and comparative interpretation of the specific results obtained from each of these constitutes probably the best scientific contribution to a field of study which presents such a high degree of complexity and uncertainty.

All the instruments have their own characteristics and usefulness in dealing with the various aspects of uncertainty. Each of the approaches proves to be particularly relevant from a certain methodological point of view, or the analysis of a specific research problem. [illegible] [illegible] in order to [illegible] [illegible] [illegible] [illegible] [illegible]

CHAPTER 7 BASIC PRINCIPLES OF ROUGH SET ANALYSIS

7.1 Introduction

The aim of this chapter is to present, in a non-technical form, the most important theoretical and applicational aspects of rough sets (Pawlak, 1982, 1991; Slowinski, 1992a). This new instrument of analysis has become well-known over the past few years and has been applied in various fields with considerable success and in particular with regard to the problem of multi-attribute sorting, discussed later in this chapter. Recently the application of the rough set theory has been suggested for the analysis of multi-attribute preference systems, in choice and ranking problematics.

7.2 Uncertainty and Rough Set Analysis

There is no doubt that it is more difficult to find a scientific approach to problems pertaining to the social sciences and, generally speaking, dealing with issues which involve some kind of uncertainty, than to those pertaining to physical sciences. In the latter case it is usually possible to obtain a precise measure of the phenomena under examination and to verify any hypotheses experimentally. In those instances where there is less possibility for making accurate measurements - and where therefore it is necessary to introduce suitable approximate scales of measurement and where the deterministic relationship between cause and effect is less evident, making it necessary to introduce stochastic models or to work on an unrealistic hypothesis of uniform behaviour - the purely scientific approach becomes more difficult. Nevertheless, such an approach is certainly not impossible, and indeed it represents a challenge. Moreover it can help to better understand a complex reality and, therefore, to offer a real contribution to the development of science and to the progress of society in general.

Uncertainty pervades every sphere of the real world. It is sometimes generated by lack of information; sometimes by the inaccuracy of available information; and sometimes by the imperfect representation or configuration of a real phenomenon.

Many attempts have been made, all of them to some degree successful, to tackle in a scientific way those problems of uncertainty caused by inaccurate information: probability calculation, credibility theory, statistical inference and fuzzy sets. To these, rough set theory has recently been added, proving to be a useful instrument for the analysis of decision situations in the presence of vague or inaccurate descriptions of reality.

The aim of the rough set approach is to pinpoint regularities in the data, in order to draw from them elements and relationships which are not immediately evident but which can be useful in a lot of analyses and of great help in a decision making problem context. More exactly, this approach attempts to discover possible cause-effect relationships between the data available, to underline the importance and the strategic role of some data and the irrelevance of other data. Moreover, this approach does not require any preliminary or further information about the data available, which may also be qualitative; it is also able to take into consideration any existing inconsistent information which is not automatically eliminated *a priori*, since it is possible to obtain facts from both positive and negative examples.

With reference to decisional aspects, rough sets can be used to advantage in problems of choice, sorting and ranking of actions in the field of multi-attribute decision support. Moreover, this approach, in building a model of the decision-maker's preferences, does not require the specification of any parameters, but simply uses examples supplied by the decision-maker, from which to obtain decisional rules which, discussed and acknowledged by the decision-maker himself, can then be used to aid the decision process.

7.3 Basic Concepts of Rough Set Analysis

With reference to a certain finite set of objects U, known as the universe of the discourse, it is assumed possible to perceive the differences between them by observing some information associated with each of them. It is, therefore, vital to identify a finite set Q of attributes which serve to identify and characterize these objects. The rough set theory, proposed in the early 1980s by Pawlak (1982), provides a formal tool for transforming a data set into knowledge, in the sense of ability to classify and distinguish data on the basis of the different values their attributes assume with reference to each object. Each attribute $q \in Q$ must, therefore, be able to assume different values in its domain U_q. There must be at least two of these values in order for the attribute to be a significant basis for the required characterization. If an attribute is quantitative, its domain is, in practice, partitioned into a suitable number of sub-intervals, which give a good description of the phenomenon studied, so as to avoid ending up with a distribution of values with a high number of modalities, not very useful for the analysis intended. The choice of the bounds, called norms, used to define these sub-intervals is an important and difficult task, to ensure a correct application of this approach and to ensure that not too much information is lost in the translation of original quantitative attribute values into qualitative coded values.

At this point, to every object $x \in U$ may be associated a vector whose components are the distinct evaluations of x with respect to every attribute of Q and called a description of x in terms of attribute-values from the set Q. The table containing the descriptions of every $x \in U$ by means of the attributes of the set Q is known as the information table. It is possible to obtain a description of $x \in U$ also in terms of any one subset of attributes $P \subseteq Q$.

A fundamental concept of the rough set theory is that of the binary relation of indiscernibility, denoted I_P. Two objects $x, y \in U$ are said to be P-indiscernible by means of the set of attributes $P \subseteq Q$ if and only if they have the same description. Naturally, the binary relation I_P is reflexive, symmetric and transitive (equivalence relation); its classes, that is the subsets of U containing all objects having the same description in terms of the attributes from subset P, and only these, are called P-elementary sets. If all the attributes of Q are considered, the Q-elementary sets are called atoms. The P-elementary sets, $P \subseteq Q$, generate a partition of U, in that every object x ∈ U belongs to one and only one P-elementary set.

Thus, for example, let U be the set of people present at an international conference, which are to be described by means of the attributes of the set Q = {sex, colour of eyes, colour of hair, height}. Let {male (M), female (F)}, {light (L), dark (D)}, {fair (F), dark (D), black (B)} be the respective domains of the first three (qualitative) attributes of Q. Let the height domain (quantitative attribute) be defined, for example, with the norms and code system depicted in Table 7.1.

Table 7.1. Norms and code system.

code	height in cm
A	$h < 160$
B	$160 \leq h < 170$
C	$170 \leq h < 180$
D	$180 \leq h < 190$
E	$h \leq 190$

If Agatha and Ann are girls, with blue eyes and green eyes respectively, both with blond hair, 166 cm and 160 cm tall, their descriptions in terms of the attributes of Q will be respectively:

Des_Q (Agatha) = {F, L, F, B}
Des_Q (Ann) = {F, L, F, B}

That is to say that Agatha and Ann are Q-indiscernible (and naturally P-indiscernible for every $P \subset Q$). If, next, Joseph has blue eyes, and blond hair and is 178 cm tall, that is:

Des_Q (Joseph) = {M, L, F, C}, and if
P = {colour of eyes, colour of hair} ⊂ Q,

Agatha and Joseph will be P-indiscernible but not Q-indiscernible.

If S = {sex} ⊂ Q, two S-elementary sets will be obtained, containing respectively all the males and all the females present at that conference. It is also clear that a modification in the domain of one or more of the attributes leads to a change in the indiscernibility relations. Thus, for example, if the domain of the "height" attribute is modified, considering a variation of only 5 cm between each level, Agatha and Ann will no longer be Q-indiscernible, although they will still be P-indiscernible and S-indiscernible.

In our example the P-elementary sets will consist of the subsets, some of which may be empty, of people having the following combinations of characteristics:
P1 : light eyes, fair hair
P2 : light eyes, dark hair
P3 : light eyes, black hair

P4 : dark eyes, fair hair
P5 : dark eyes, dark hair
P6 : dark eyes, black hair.

Observe that 6 potential P-elementary sets are obtained, that is 6 = card (domain of "colour of eyes") x card (domain of "colour of hair") = 2 x 3, where card (X) indicates the number of objects belonging to the (finite) set X. In general therefore, the maximum number of atoms ($U \neq \emptyset$) is

$$0 < \prod_{q \in U} \text{card}\ (U_q) \leq \text{card}\ (U) \tag{7.1}$$

where card (U) is the number of elements in set U (in the example originally proposed the result is: 2 x 2 x 3 x 5 = 60). This number, therefore, increases with an increase in the number of significant attributes considered and, for an equal number of attributes with the cardinal of the domain of each attribute. It is equal to the card (U). That is, each atom contains only one element if and only if the combination of attributes and domains is totally discriminating (selective information table). This is not always possible, nor is it always desirable since it would not lead to information useful for the analysis.

Recently, as an alternative to the indiscernibility relation described here, a proposal has been made to introduce a similarity relation, weaker than the previous one, which is only reflexive; it may not be either symmetric or transitive. The use of this relation may be convenient in particular when quantitative attributes are used, because it would reduce some problems due to the arbitrary introduction of norms for the subdivision of the domain of an attribute into suitable sub-intervals. The asymmetry of the similarity relation is due to the fact that one object is considered as the reference and another as the subject of the same. Thus, for example, if two people are considered similar with respect to the height attribute, the difference between their heights being less than 10 per cent, where the height of a is 165 cm and that of b 183 cm, we obtain that "b is similar to a" (because $183 - 165 = 18 < 0.10 \times 183$), but at the same time "a is not similar to b" (because $18 > 0.10 \times 165 = 16.5$).

The P-indiscernibility relation and the equivalence classes generated by this relation constitute respectively the basic concept and the bricks used to build rough set theory.

For the definition of rough sets it is necessary to introduce two other key-concepts. Let $P \subseteq Q$ be a subset of attributes and $X \subseteq U$ a subset of objects of U. We define as P-lower approximation of X, denoted with $P_L X$, the subset of U having as its elements all objects belonging to the P-elementary sets contained in the set X, and only these. In other words, the elements of $P_L X$ are all elements of U belonging to all classes generated by the indiscernibility relation I_P and contained in X, and only these.

We define as the P-upper approximation of X, denoted with $P_U X$, the subset of U having as its elements all objects belonging to the P-elementary sets having at least one element in common with the set X, and only these. In other words, the elements of P_UX are all the elements of U belonging to all classes generated by the indiscernibility relation I_P which have at least one representative belonging to X, and only these.

The difference between these sets is known as P-boundary of X, denoted by Bn_P (X):

$$Bn_P (X) = P_UX - P_LX \tag{7.2}$$

Therefore, $P_LX \subseteq X \subseteq P_UX$ results and if an object x belongs to P_LX it is certainly also an element of X; if x belongs to P_UX it may belong to the set X; Bn_P (X) therefore constitutes the doubtful region: with reference to its elements nothing can be said with certainty about its belonging to the set X. The indiscernibility classes generated by I_P therefore constitute the basic instrument of the rough set theory used to obtain a better knowledge of reality. This knowledge, therefore, is intended as a family of partitions of U, generated by the indiscernibility relation I_P on U, $P \subseteq Q$.

A P-rough set is the family of all subsets of U which have the same lower and upper P-approximations. A rough set, therefore, does not have associated to it a characteristic function; this usually is $\mu_X : U \rightarrow \{0, 1\}$ in the classic set theory (where $\mu_X (x) = 1$ if and only if x belongs to X, or $\mu_X (x) = 0$, if and only if x does not belong to X) or $\mu_X: U \rightarrow [0, 1]$ in the fuzzy set theory (where $0 \leq \mu_X (x) \leq 1$ is a membership function indicating exactly the degree of belonging of an element $x \in U$ to a certain set X). In the rough set theory the intention is to approximate a set X, $X \subseteq U$, by means of a pair of sets associated to it, called lower approximation, P_LX, and upper approximation P_UX, of X, that can be then considered as a particular case of the interval set. Only if $P_UX = P_LX$, does X prove to be equal to the union of a certain number of P-elementary sets and is called P-definable. Clearly, in this case (and only in this case) it is possible to affirm with certainty whether x, $x \in U$, belongs to X, $X \subseteq U$ or not, by using the set of attributes P. Moreover, the accuracy of the approximation of X, equal to $\frac{\text{card } (P_LX)}{\text{card } (P_UX)}$, will then be at its maximum value (that is equal to one). In general, therefore, the aim of the rough set analysis is to establish whether x is an element of X on the basis of the lower and upper approximations of X, rather than directly by means of a specific characteristic function.

This approach, therefore, leads to an imprecise representation of reality due to the 'granularity' of knowledge, a term introduced by Pawlak (1982). In other words, reality is represented by granules, corresponding to the P-elementary sets, that is to subsets of U whose elements are indiscernible, undistinguishable, by the set of attributes $P \subseteq Q$ used, because they present the same description in terms of the value of these attributes. The granularity of knowledge representation is used to define the key-concepts of the rough set theory. The size of these granules depends, naturally, both on the number of attributes used for the description of the objects in the original information table, and on the domain of each attribute. With a suitable variation in these two quantities it is possible to obtain a variation in the dimensions of the granules: an increase in the number of attributes belonging to the set P and in the number of values which each attribute can assume, does not result in an increase in the granularity, that is in the dimension of the P-elementary sets. In other words, the granularity should increase, but this is not a sure effect of the mentioned changes.

Therefore, the lower and upper approximations of X are determined on the basis of a typology (granules) generated by the available information on the elements of the set (indiscernibility relation), that is on the ability to perceive some real phenomena, classify them, distinguish them on the basis of the information obtained from an observation of them or on the basis of the knowledge of them possessed by an agent, an expert. The representation of reality by means of rough sets is therefore based on the knowledge possessed of reality and the capacity to classify the information obtained.

A well-known example may clarify this concept somewhat further. With reference to the representation of pictures (photography, television, etc.) it may be found that the passage from black-and-white representation to that which uses different levels of grey is similar to the difference between using traditional sets and using fuzzy sets. In the same way, the dimensions of the pixels, that is the greater or smaller definition of the picture, is the consequence of a representation of reality by means of rough sets.

7.4 Reduction and Dependence of Attributes

Let $\mathbf{Y} = (Y_1, Y_2, ..., Y_n)$ be a partition of U built independently on the evaluations of the objects of U with respect to the attributes from set $P \subseteq Q$ (for example, the evaluation of an impartial expert or an agent, or results of past studies, or examples of real decisions). With reference to the partition **Y**, we denote as P-lower approximation and P-upper approximation respectively the sets having as their elements the P-lower and P-upper approximations of its classes, that is $P_L\mathbf{Y} = (P_LY_1, P_LY_2, ..., P_LY_n)$ and $P_U\mathbf{Y} = (P_UY_1, P_UY_2, ..., P_UY_n)$. An indicator of the quality of the approximation of the partition **Y** by means of the set of attributes P, notation $\gamma_P(\mathbf{Y})$, is given by the ratio between the total number of P-correctly classified objects (that is, belonging to the P-lower approximations of Y_i, i = 1, 2, ..., n), and the total number of objects considered; that is, the quality of the classification is:

$$\gamma_P(Y) = \frac{\sum_{i=1}^{n} \text{card}(P_L Y_i)}{\sum_{i=1}^{n} \text{card}(U)} \tag{7.3}$$

This index assumes its maximum value of one if and only if each of the classes Y_i of **Y** prove P-definable, that is if each of them is given by the union of P-elementary sets.

A first interesting application of rough set analysis, which can be made with reference to the quality of the classification mentioned above, regards the dependency and reduction of attributes.

It is said that the set of attributes $R \subseteq Q$ depends on the set of attributes $P \subseteq Q$, notation $P \rightarrow R$, if and only if each equivalence class of the indiscernibility relation generated by P is included in some equivalence class of the indiscernibility relation generated by R, that is $P \rightarrow R$ if $I_P \subseteq I_R$. In other words, R depends on P if the classification obtained by the use of the attributes of P is more precise, more detailed, than that obtained using the attributes of R. This means that the values of the attributes in R may be uniquely determined by the values of the attributes in P.

It follows that if $P \rightarrow R$, then $P \rightarrow q$ for every attribute q belonging to R (elementary dependence). In general, there is no set relationship (i.e. intersection and/or inclusion) between the sets P and R. Consider, for example, the following table (Table 7.2), where ten participants in an international meeting are considered, classified on the basis of the following attributes of the set Q: {sex (q_1), colour of eyes (q_2), colour of hair (q_3), height (q_4)}, whose respective domains are {male (M), female (F)}, {light (L), dark (D)}, {fair (F), dark (D), black(B)}, {tall (T), medium (M), short (S)}:

Table 7.2. Coded data table.

U	q_1	q_2	q_3	q_4
x_1	F	L	F	T
x_2	F	L	F	T
x_3	F	L	F	T
x_4	F	D	B	M
x_5	F	D	B	M
x_6	M	L	L	T
x_7	M	L	D	M
x_8	M	L	D	M
x_9	M	D	D	S
x_{10}	M	D	B	T

The following dependencies between sets of attributes exist:

$\{q_1, q_3, q_4\} \rightarrow \{q_2\}$
$\{q_3, q_4\} \rightarrow \{q_1, q_3, q_4\}$
$\{q_3, q_4\} \rightarrow \{q_1\}$ (elementary dependence)
$\{q_3, q_4\} \rightarrow \{q_2\}$.

The dependency can be interpreted as follows: the $\{q_3, q_4\}$ -elementary sets, that is the sets having the same description using the attributes q_3 and q_4 ($\{x_1, x_2, x_3\}$, $\{x_4, x_5\}$, $\{x_6\}$, $\{x_7, x_8\}$, $\{x_9\}$, $\{x_{10}\}$) are all included in the q_2 -elementary sets, which are $\{x_1, x_2, x_3, x_6, x_7, x_8\}$ and $\{x_4, x_5, x_9, x_{10}\}$. In other words, the classification of the objects of U by means of the attributes of the set $\{q_3, q_4\}$ is more accurate, more precise, than the one obtainable using the attribute q_2, that is, it presents finer grains, a lower granularity.

The search for dependencies between attributes is of considerable importance in the analysis of the information table using the rough set approach.

Another fundamental concept is that of attribute reduction, that means, given a classification Y of the objects of U, the search for a minimal set of (independent) attributes R which supplies the same quality of classification as the original set of attributes P. The minimal subset $R \subseteq P \subseteq Q$ such that $\gamma_R (Y) = \gamma_P (Y)$ is called Y-reduct of P and denoted $RED_Y (P)$. Note that a single information table may have more than one reduct. The intersection of all Y-reducts is known as the Y-core of P, that is $CORE_Y (P) = \cap RED_Y (P)$. The core contains all the attributes from P which are considered of the greatest importance in the information table, that is to say the most relevant for a correct classification of the objects of U. So, for example, if we consider the classification Y = {Y1, Y2, Y3, Y4}, where Y1 = {x1, x2, x3, x4, x5}, Y2 = {x6}, Y3 = {x7, x8, x9}, Y4 = {x10}, we have three different reducts: RED^1 (Q) = {q1, q3}, RED^2 (Q) = {q2, q4}, RED^3 (Q) = {q3, q4}.

In other words, to analyse the information table, it is sufficient to use any one of the attribute reducts $R \subseteq Q$, that is the classification Y of the objects of U may be characterized without losing any information using only the attributes from R, while the information supplied by the attributes of Q - R prove redundant for this purpose. Vice versa, none of the attributes belonging to the core may be neglected without deteriorating the quality of the classification considered, that is if any one attribute belonging to the core is eliminated from the information table, it will certainly not be possible to obtain with the remaining attributes the highest quality of approximation otherwise obtainable.

The importance of knowing the reducts of Q for practical purposes is clear. Often the acquisition of evaluations by means of a certain attribute is very difficult or a very expensive operation; these evaluations need, therefore, not be carried out at all, with a great saving of resources and time, if the corresponding attributes do not belong to an effectively used reduct. In the case of the presence of more than one reduct of Q, it is therefore possible to choose from among these the most efficient one with reference to the acquisition of corresponding information. This observation is exclusively valid with reference to all data contained in the information table. Nothing can exclude *a priori* that when by considering new objects it may be possible to obtain attribute reductions which are totally or partially different from those found previously. It cannot therefore be considered methodologically correct to limit the use of the attributes to those of the latter alone, if it is necessary to analyse new objects, or new examples which may be different from those previously considered in the information table.

Finally, there exists an important relation between reducts and dependency of attributes. If R is a reduct of Q, then the set of attributes belonging to the complement of R with respect to Q depends on the attributes belonging to R, that is if R is a reduct of Q, then $R \rightarrow Q - R$. This relation is known as basic dependency between attributes. In other words, once the values assumed by the attributes of R are known, the values of the attributes of Q - R are unequivocally determined.

It is possible from the above to develop a clear understanding of the operational importance of a rough set approach: the possibility to discover all dependencies between attributes existing in an information table and, therefore, to take into consideration only the more important attributes in the analysis of the information table itself.

7.5 Decision Rules

A classification **Y** of the objects of U, that is a partition of U into the classes Y_1, Y_2, ..., Y_n, carried out by an expert or based on past experience, may always be interpreted as deduced from the binary indiscernibility relation I_D with relation to a set of attributes D. In an information table, therefore, it is possible to distinguish under this profile two non-empty classes of attributes C and D, known respectively as condition and decision attributes, such that $Q = C \cup D$ and $C \cap D = \varnothing$, which generate respectively the indiscernibility relations I_C and I_D. The corresponding table of data is called decision table and in this we will therefore find two partitions of U, the first $\mathbf{X} = \{X_1, X_2, ..., X_k\}$, whose classes are the C-elementary sets and are called condition classes; and the second $\mathbf{Y} = \{Y_1, Y_2, ..., Y_n\}$, whose classes are the D-elementary sets and are called decision classes. Hence, every object $x \in U$ is described using the different values of the condition attributes (with all C-indiscernible objects forming the different classes X_i, i = 1, 2, ..., k) and is also classified using the different values of the decision attributes (or single attribute), that is, in a manner absolutely independent of the set C of condition attributes (with all D-indiscernible objects constituting the different classes Y_j, j = 1, 2, ..., n).

An implication relation between the description of a condition class X_i and the description of a decision class Y_j, that is $Des_C(X_i) \rightarrow Des_D(Y_j)$, is called a (C,D)-decision rule. For each class Y_j, j = 1, 2, ..., n, the set of decision rules is;

$$\{r_{ij}\} = \{Des_C(X_i) \rightarrow Des_D(Y_j) : X_i \cap Y_j \neq \varnothing, i = 1, 2, ..., k\} \tag{7.4}$$

The rule r_{ij} is exact or deterministic, if and only if $X \subseteq Y$; otherwise, r_{ij} is approximate or non-deterministic. An exact rule r_{ij} guarantees that the values of the decision attributes that characterize the class Y_j correspond unequivocally to some combination of values of the condition attributes, which characterize the class X_i (same conditions, same decisions). An approximate rule, on the other hand, states that more than one value of the decision attributes corresponds to the same combination of values of the condition attributes (same conditions, different decisions). In the case of exact rules, therefore, using the information contained in the decision table it is always possible to state with certainty whether or not an object $x \in X_i$, i = 1, 2, ..., k, belongs to the class Y_j, j = 1, 2, ..., n. On the other hand, the approximate description of reality as a consequence of the granularity of knowledge makes it impossible, in the case of approximate rules, by using the information available in the decision table, to decide whether or not an object x belonging to the boundary of X_i, i = 1, 2, ..., k, belongs or not to a given decision class Y_j, j = 1, 2, ..., n.

An exact rule, thus, offers a sufficient condition of belonging to a decision class; an approximate rule admits the possibility of this.

The decision rules may also be expressed with logical statements "if ...., then ...", relating descriptions of condition and decision classes. This way of expressing the decision rules does not only have the advantage of being perfectly comprehensible to the decision maker, but also that of being the simplest and most natural way to build a preference model in the light of a sufficiently wide sample of illustrative comparisons. In multi-attribute sorting problems, the preferential information consists of sorting examples, that is in the assignment to particular decision categories of actions described by particular condition attribute-values. The rules derived from examples enable the decision maker to understand the reasons of his/her preference, while their recognition justifies their use for decision support.

From this point of view, therefore, it is possible to verify the information contained in the original decision table and to enrich it, acquiring further decision rules directly by means of suitable interviews or discussions with experts. Thus, it is possible to acquire information also directly in the form of decision rules supplied by experts, thereby enriching the original information contained in the decision table.

In practice, it is possible to use both the decision rules obtained by elaborating the data contained in the decision table and, if necessary, further rules supplied by experts. The former may be accompanied by an indicator of their "strength", that is, for example, the frequency (absolute or relative) of examples in agreement with each decision rule. Moreover, both the former and the latter may be based on suitable and different sets of condition attributes, containing a greater or smaller number of attributes, even a single attribute. This final case implies that the value k assumed by the attribute C_j is sufficient to guarantee that the decision attribute (or attributes) will assume certain values, whatever the values of the other condition attributes. This consideration assumes particular importance when incomplete information is used. That is when some values in the decision table are missing or are uncertain. Decision rules which do not use condition attributes containing imperfect information assume particular importance in such cases, since they make it possible to operate without the knowledge of these values.

7.6 Multi-attribute Choice, Sorting and Ranking

Decision rules, which constitute one of the most relevant aspects of rough set analysis, may be applied immediately in order to offer recommendations and advice in problems of multi-attribute sorting, that is in the assignment of each potential action to an appropriate pre-defined category according to particular aims. In this case the classification of a new object may be usefully aided by a comparison between its description and the values contained in the decision rules. These are more general than the information contained in the original decision table and permit a classification of new examples in a larger number and more easily than would be possible by using a direct comparison between these and the original examples. From the comparison with the decision rules one of the following possibilities, in general, may occur:

a) the new object matches an exact rule;

b) the new object matches more than one exact rule, indicating the same decision class;

c) the new object matches an approximate rule or more than one rule, indicating different decision classes;

d) the new object does not match any decision rule.

In the first two cases, the classification of the new object presents no difficulty. In case c), where some ambiguity exists, it may be possible to improve the strength of the original examples to which each of the suggested decision classes corresponds, in order to assist the user more properly. In case d), where there is no rule matching the description of the new object, the decision maker may be helped by presenting him with the set of rules nearest to that description; this may be done by introducing, for example, a distance measure, based on a valued closeness relation R, which, by introducing indifference, strict difference and veto thresholds on particular attributes, makes use of suitable concordance and discordance tests.

The rough set approach has recently also been applied to multi-attribute choice and ranking problems. In this case the original information table must be modified in such a way as to consent suitable pairwise comparisons. It is then possible to proceed with the construction of a suitable pairwise comparison table (PCT) as follows.

In each line the object under consideration is a specific, ordered pair of actions (a,b) belonging to a suitable sample B (non-empty) of comparisons, $B \subseteq A \times A$, where A is a set of actions, chosen by the decision maker as a basis on which to carry out the pairwise comparisons. Correspondingly, the description of each pair (a,b) by means of the condition attributes will be given by suitable binary relations, exactly one for each attribute or, if possible, relevant criterion (i.e. an attribute whose domain is ordered according to decreasing or increasing preference).

Finally, when a set of binary relations on the set A has been defined, indicating comprehensive pairwise comparisons, one and only one of these binary relations is associated to every pair $(a,b) \in B$, which induces, therefore, a partition of B and which assume the same role as a decision attribute. The PCT introduced may, therefore, be considered a decision table, since it may contain distinct condition and decision attributes.

The preferential information is therefore represented by associating, to every pair (a,b) of actions of the sample of comparisons, binary relations which may for example express the degree of preference or indifference of a over b with respect to each condition attribute considered and the degree to which a is comprehensively preferred to b. Naturally, the representation of the pairwise comparisons is strictly correlated to the ordering properties of preference relations defined on each attribute (or criterion). In particular, in the presence of a criterion it may also be possible to measure the strength of positive or negative preference of a over b, which then - utilising suitable thresholds - may be transformed into the specific binary relations mentioned above.

With reference to the PCT defined it is then possible to adapt all principal concepts typical of rough set analysis (indiscernibility relation, reduction of attributes, core, decision rules, etc.); in particular, it is possible to define directly a rough binary relation.

In order to solve multi-attribute choice and ranking problems it is necessary to exploit the set of decision rules obtained, which, after being approved by the decision-maker, become his personal preference model. This model is then used to obtain binary preference relations with reference to a wider set of potential actions, containing also the new actions among which the choice or ranking is to be made. To this purpose, therefore, the descriptions of each new pair of actions are compared with those obtained from the decision rules relative to the chosen sample of B. Also in this case, the comparison may result in one of the four situations, i.e. the correspondence, the lower approximation of a global preference relation (cases a and b), and the upper approximation of the same binary relation (cases a, b or c). Finally, these rough binary relations may then be used in a suitable way and applied to the previous techniques, in line with the rough approach adopted and with techniques and concepts belonging to interval sets in order to obtain a choice result.

The approach which utilizes the preferential information contained in a PCT can also easily be extended to multi-attribute sorting problems, in a complementary manner with respect to the original rough set analysis.

7.7 Relevance of Rough Set Analysis for Meta-analytical Research

The nice characteristics of rough set theory - and especially its ability to consider qualitative data, to create a proper reduction of attributes, to build preference models in terms of "if ..., then ..." rules, and to be relevant for traditional multi-attribute decision problems - make this approach an extremely interesting tool for application in meta-analysis.

This technique may be useful in many phases of a meta-analytic study, such as, for example, in choice of individual studies to be analysed and the minimal description of a concept, in addition to the actual study phase of the synthesis. The technique may be a very interesting complement - not an alternative - to the statistical techniques normally adopted, since it has a wider field of application than classical multiple regression analysis, and while it also has greater flexibility with regard to both input data and information it can supply. Not only is it possible to treat qualitative data in a 'natural' way, thus obtaining the corresponding objective information, but this method can be elaborated to cases which contain partial information. The possibility, moreover, of enriching the initial information table, by means of the direct acquisition of new knowledge in the form of decision rules for experts, constitutes a characteristic of significant potential in the field of meta-analysis.

The information which may be obtained is also easy to interprete for the decision-maker, since it is in the form of "if ..., then ..." rules. In particular, with reference to multi-attribute or multicriteria problems, recent extensions of the rough set approach make it also applicable to problems of choice and ranking, which are probably among the most recurrent and interesting issues encountered in meta-analysis for environmental economics.

CHAPTER 8 TECHNICAL ISSUES IN ROUGH SET ANALYSIS

8.1 Introduction

This chapter provides a technical background to rough set analysis, the basic principles of which were outlined in Chapter 7. The technicalities of the method are set out in Section 8.2. This is followed by a reflection on the management of thresholds in Section 8.3. In addition, the chapter evaluates the advantages of rough set analysis in comparison with more traditional ways of dealing with incomplete data sets (Sections 8.4 and 8.5) in order to support the use of this method in several of the case studies set out in later chapters of the book. The advantages of the rough set approach compared with the fuzzy set approach are reviewed in Section 8.4, while Section 8.5 deals with the merits of the rough set approach relative to the traditional method of statistical analysis.

8.2 Technical Background of Rough Set Analysis

8.2.1 Information table and indiscernibility relations

An information table is a finite data table with rows corresponding to objects and columns to attributes; its entries are attribute values. Thus an information table can be considered as a collection of objects described by values of attributes, representing the properties of the objects (Mrozek, 1985, 1989). Then an information table is formally the 4-tuple S=(U,Q,V,f), where:

U is a finite (non empty) set of objects, called universe;
Q is a finite set of attributes;
$V= \cup_{q \in Q} V_q$, and V_q is the domain of the attribute q;
$f : UxQ \rightarrow V$ is a total function such that $f(x,q) \in V_q$ for every $x \in U$, $q \in Q$, called information function.

Any pair (x,f(x, q)), $x \in U$, $q \in Q$ is called descriptor in S.

Let $P \subseteq Q$ and $x,y \in U$. Then x and y are indiscernible by a set of attributes P in S (denotation xIND(P)y) iff f(x,q) = f(y,q), for every $q \in P$; that is, objects x and y are indiscernible in terms of considered attributes. If any attribute q has a quantitative nature, its domain V_q is, in practice, partitioned into m_q subintervals by means of suitable bounds, called norms. In this case x and y are indiscernible by a set of attributes P in S iff f(x,q) and f(y,q) are identical or belong to the same subinterval, for every $q \in P$. After the translation of original quantitative attribute-values in qualitative coded values, we speak of coded data table. The definition of the norms is arbitrary and a difficult task, because it may influence the results of the rough set analysis.

The binary indiscernibility relation IND(P) is an equivalence relation on U for every $P \subseteq Q$. Equivalence classes of relation IND(P) are called P-elementary sets in S and correspond to granules of knowledge representation. Q-elementary sets are called atoms of S (f(x,q) = f(y,q), for every $q \in Q$). Any finite union of P-elementary sets is called a P-definable set in S. Information table S is selective iff all atoms in S are one-element sets. The family of all equivalence classes of relation IND(P) on U is denoted by U|IND(P) or U|P; $[x]_P$ denotes the equivalence class of IND(P) containing object $x \in U$.

Example 8.1

Let U = {x1, x2, ..., x10}, Q' = {c1, c2, d} with values shown in Table 8.1.

Table 8.1. Values of quantitative attributes in Example 8.1.

U	c_1	c_2	d
x1	0	1280	0
x2	0	1320	0
x3	0	1480	0
x4	0	2600	1
x5	0	3100	1
x6	1	700	0
x7	1	1120	0
x8	1	1260	0
x9	1	1450	1
x10	1	2200	1

Norms of attribute c2: 1000, 1500;

original value of c2	coded value of c2
$x \leq 1000$	L
$1000 < x \leq 1500$	M
$1500 < x$	H

Table 8.2. Coded data table.

U	c_1	c_2	d
x1	0	M	0
x2	0	M	0
x3	0	M	0
x4	0	H	1
x5	0	H	1
x6	1	L	0
x7	1	M	0
x8	1	M	0
x9	1	M	1
x10	1	H	1

{c1, c2}-elementary sets:
{x1, x2, x3}, {x4, x5}, {x6}, {x7, x8, x9}, {x10};
{c2}-elementary sets:
{x6}, {x1, x2, x3, x7, x8, x9}, {x4, x5, x10};
{d}- elementary sets:
{x1, x2, x3, x6, x7, x8}, {x4, x5, x9, x10};
atoms (Q'-elementary sets):
{x1, x2, x3}, {x4, x5}, {x6}, {x7, x8}, {x9}, {x10}.

$Des_P(x)$ denotes the description of object $x \in U$ in terms of attribute-values from subset P, i.e.

$$Des_P(x) = \{(q,v) : f(x,q) = v, \forall q \in P\} \tag{8.1}$$

Since all objects belonging to the same equivalence class (P-elementary set) are indiscernible, it is clear that $Des_P([x]_P) = Des_P(x)$. According to the definition (8.1), description is formally a set of attribute-value pairs. Sometimes it is more natural to understand description as a linguistic conjunction of attribute-value pairs.

8.2.2 Approximation of sets and rough sets

Let X be a subset of U, R an indiscernibility relation on U and $[x]_R$ the equivalence class of R containing object $x \in U$. We call the ordered pair $A = (U,R)$ the approximation space. In order to define X in terms of definable sets (finite union of equivalence classes of R) in A, we introduce the concepts of lower and upper approximation of a set (Pawlak and Slowinski, 1993; Ziarko, 1993). The lower approximation of X in A, denoted R_LX, is the greatest definable set in A, contained in X, i.e.

$$R_LX = \{x \in U : [x]_R \subseteq X\} \tag{8.2}$$

The upper approximation of X in A, denoted R_UX, is the least definable set in A, containing X, i.e.

$$R_UX = \{x \in U : [x]_R \cap X \neq \emptyset\} \tag{8.3}$$

A rough set in A is the family of all subset of U having the same lower and upper approximations in A. The difference between the upper and the lower approximation of X in A is called boundary of X in A (doubtful region), denoted $Bn_R(X)$, i.e.

$$Bn_R(X) = R_UX - R_LX \tag{8.4}$$

Let X and Y be subsets of U; lower and upper approximations in A have among others the following properties:

$$R_LX \subseteq X \subseteq R_UX; \tag{8.5}$$
$$R_LU = R_UU = U; \tag{8.6}$$
$$R_L\varnothing = R_U\varnothing = \varnothing; \tag{8.7}$$
$$X \subseteq Y \Rightarrow \begin{cases} R_LX \subseteq R_LY \\ R_UX \subseteq R_UY; \end{cases} \tag{8.8}$$

$$R_L(X \cup Y) \supseteq R_LX \cup R_LY; \tag{8.9}$$
$$R_U(X \cup Y) = R_UX \cup R_UY; \tag{8.10}$$
$$R_L(X \cap Y) = R_LX \cap R_LY; \tag{8.11}$$
$$R_U(X \cap Y) \subseteq R_UX \cap R_UY. \tag{8.12}$$

Let x be an object of U; we say that x is certainly an element of X iff $x \in R_LX$, and that x is possibly an element of X if $x \in R_UX$. This is because we want to decide if x is in X on the basis of definable sets in A, rather than directly on the basis of X. Since $R_LX \subseteq X \subseteq R_UX$, if x is in R_LX, it is certainly in X, and if x is in R_UX, it is possibly in X.

If X is definable in A, we know exactly whether any object belongs or does not belong to X, using only the lower approximations of X in A; otherwise we only know that approximately. We can consider the following types of approximation:

if $R_LX \neq \varnothing$ and $R_UX \neq U$, X is roughly definable in A;
if $R_LX = \varnothing$ and $R_UX \neq U$, X is internally non-definable in A;
if $R_LX \neq \varnothing$ and $R_UX = U$, X is externally non-definable in A;
if $R_LX = \varnothing$ and $R_UX = U$, X is totally non definable in A.

This is because if $R_LX = \varnothing$ or $R_UX = U$, we have no useful information about lower or upper approximation of X in A.

Let now $P \subseteq Q$ and $Des_P(X)$ be the description of equivalence class X (P-elementary set), $X \in U|P$, i.e. $Des_P(X) = \{(q,v) : f(x,q) = v, \forall x \in X, q \in P\}$. In order to evaluate how well the set $\{Des_P(X) : X \in U|P\}$ describes objects of any set $Y \subseteq U$, we can use the following concepts:

the P-lower approximation of Y in S, denoted by P_LY and defined as:
$$P_LY = \cup \{X \in U|P : X \subseteq Y\}; \tag{8.13}$$

the P-upper approximation of Y in S, denoted by P_UY and defined as:
$$P_UY = \cup \{X \in U|P : X \cap Y \neq \varnothing\}; \tag{8.14}$$

the P-boundary of Y in S, denoted by $Bn_P(Y)$, defined as:
$$Bn_P(Y) = P_UY - P_LY. \tag{8.15}$$

Set P_LY and set P_UY are the set of all objects belonging to U which can be respectively certainly or possibly classified as elements of Y, using the set of attributes P only. The set $Bn_p(Y)$ is the set of objects which cannot be certainly classified on the basis of descriptions of the P-elementary sets.

Given a lower and an upper approximation of a set $Y \subseteq U$, $(y \neq \varnothing)$ it is possible to define a measure of the accuracy and a measure of the quality of these approximations, in order to define how exactly one can describe the considered set of objects using the available information from the set of attributes $P \subseteq Q$. We can then define the accuracy of approximation of set Y by P in S (shortly, accuracy of Y), given by:

$$\alpha_P(Y) = \frac{card(P_L Y)}{card(P_U Y)} \quad (8.16)$$

that expresses the ratio of all objects certainly belonging to Y and all objects possibly belonging to Y; of course, $0 \leq \alpha_P(Y) \leq 1$, and $\alpha_P(Y) = 1$ if set Y is P-definable in S; the quality of approximation of set Y by P in S (shortly, quality of Y), given by:

$$\gamma_P(Y) = \frac{card(P_L Y)}{card(Y)} \quad (8.17)$$

that expresses the percentage of P-correctly classified objects of Y.

Example 8.2

Consider the information table of Example 8.1 and let C' = {c2} and Y1 = {x4, x5, x9, x10}. We have the following approximations:

C_L'Y1 = {x4, x5, x10};
C_U'Y1 = {x1, x2, x3, x4, x5, x7, x8, x9, x10};
$Bn_{C'}$ (Y1) = C_U'Y1 - C_L'Y1 = {x1, x2, x3, x7, x8, x9};
$\alpha_{C'}$ (Y1) = 3/9 = 0.3.
Now let C'' = {c1, c2}. We obtain the following approximations:
C_L''Y1 = {x4,x5} ∪ {x10} = {x4, x5, x10};
C_U''Y1 = {x4, x5} ∪ {x7, x8, x9} ∪ {x10} = {x4, x5, x7, x8, x9, x10};
Bn_C (Y1) = C_U''Y1 - C_L''Y1 = {x7, x8, x9};
α_C (Y1) = 3/6 = 0.5.

That is, the lower approximation of class Y1 using attributes c1 and c2 is the set {x4, x5, x10}, included in Y1 and containing all objects belonging to the same equivalence classes, described by {0, H} and {1, H}; the upper approximation {x4, x5, x7, x8, x9, x10} is the least union of {c1, c2}, equivalence classes, described by {0, H}, {1, M} and {1, H} - including Y1.

Let be S1 = {x4, x5, x7, x10}, S2 = {x4, x5, x8, x10},
S3 = {x4, x5, x9, x10},
S4 = {x4, x5, x7, x8, x10}, S5 ={ x4, x5, x7, x9, x10},
S6 = {x4, x5, x8, x9, x10}.

Table 8.3. Equivalence classes with respect to attributes c1 and c2 and approximations of set Y1.

		c2 L	c2 M	c2 H
c1	0		x1, x2, x3	x4, x5
	1	x6	x7, x8, x9	x10

The family {S1, S2, S3, S4, S5, S6} is a rough set in the space A = {U, IND(C'')} (because C_L''(S1) = C_L''(S2) = ... = C_L''(S6) and C_U''(S1) = C_U''(S2) = ... = C_U''(S6)). Set Z1 = {x4, x5, x6} is definable in A Z1 = {x4, x5} ∪ {x6}; set Z2 = {x3, x4, x6, x7, x10} is externally non-definable in A (C_U''(Z1) = U) and set Z3 = {x3, x4} is internally non-definable in A (C_U''(Z3) = ∅).

Let **Y** = {Y_1 , Y_2 , ..., Y_n } be a partition of U, and Y_i its classes; the origin of this partition is independent on attributes from $P \subseteq Q$; the classes can be sorting examples constructed by an agent, or examples of real decisions, or results of studies made in the past.

If every class Y_i of **Y** is P-definable, then classification **Y** is called P-definable. We define P-lower and P-upper approximation of **Y** in S the sets $P_L\mathbf{Y} = \{P_LY_1 , P_LY_2 , ..., P_LY_n \}$ and $P_U\mathbf{Y} = \{P_UY_1 , P_UY_2 , ..., P_UY_n \}$ respectively.

In order to evaluate how well the set of attributes $P \subseteq Q$ approximates a given partition **Y**, the following measures are used:

quality of approximation of **Y** by **P** in **S**, or, in short, quality of classification, denoted by $\gamma_P(\mathbf{Y})$ and defined as

$$\gamma_P(Y) = \frac{\sum_{i=1}^{n} \text{card}(P_L Y_i)}{\text{card}(U)} \tag{8.18}$$

that expresses the ratio of all P-correctly classified objects to all objects in the system;

accuracy of approximation of **Y** by **P** in **S**, or, in short, accuracy of classification, denoted by $\alpha_P(\mathbf{Y})$ and defined as

$$\alpha_P(Y) = \frac{\sum_{i=1}^{n} \text{card}(P_L Y_i)}{\sum_{i=1}^{n} \text{card}(P_U Y_i)} \tag{8.19}$$

It is $0 \le \alpha_P(\mathbf{Y}) \le \gamma_P(\mathbf{Y}) \le 1$.

Example 8.3

With respect to the information table of Example 9.1, let **Y**= {Y1, Y2} be a partition of U, where Y1 = {x4, x5, x9, x10}, Y2 = {x1, x2, x3, x6, x7, x8} and C'' = {c1, c2}. We have:
C_L''Y1 = {x4, x5} ∪ {x10};
C_U''Y1 = {x4, x5,} ∪ {x7, x8, x9} ∪ {x10};
C_L''Y2 = {x1, x2, x3} ∪ {x6};
C_U''Y2 = {x1, x2, x3} ∪ {x6} ∪ {x7, x8, x9}.

Quality of approximation: $\gamma_{C''}(\mathbf{Y}) = 7/10$;
Accuracy of approximation: $\alpha_{C''}(\mathbf{Y}) = 7/13$.

8.2.3 Reduction and dependency of attributes

Some important issues of the rough set approach to knowledge analysis involve attribute reduction, such that the reduced set of attributes provides the same quality of classification as the original one, and discovering dependencies among attributes (Pawlak and Slowinski, 1993; Slowinski, 1992b).

It is said that the set of attributes $R \subseteq Q$ depends on the set of attributes $P \subseteq Q$ in S, denoted $P \rightarrow R$, if and only if each equivalence class of the indiscernibility relation generated by P is included in some equivalence class of the indiscernibility relation generated by R, i.e.

$$P \rightarrow R \text{ iff } IND(P) \subseteq IND(R) \tag{8.20}$$

Intuitively, R depends on P if the classification U|P is more "precise" than that by U|R, that is if there is a functional dependency between values of R and P, in the sense that values of attributes in R are uniquely determined by values of attributes in P. Let $P \subseteq Q$, card (P) ≥ 2. Set P is dependent in S if there exists a set $P' \subset P$ such that IND(P') = IND(P). Set $P \subseteq Q$ is a reduct of Q in S if P is an independent set in S and U|P = U|Q. Let **Y** be a partition of U; the independent minimal subset of attributes which ensures the same quality of classification as the whole set Q is called reduct of Q.

Given the partition **Y** of U, the minimal subset $R \subseteq P \subseteq Q$ such that $\gamma_P(\mathbf{Y}) = \gamma_R(\mathbf{Y})$ is called **Y**-reduct of P (or reduct of P) and denoted by RED (P). An information table S may have more than one reduct. The intersection of all **Y**-reducts is called the core of P, i.e.

$$CORE_Y(P) = \cap RED_Y(P) \tag{8.21}$$

The core is an important issue in the rough set approach to sorting problems: it is a collection of the most significant attributes from Q in the system.

An important property that establishes relationship between reducts and dependency is: if R' is a reduct of R, then $R' \rightarrow R - R'$.

This kind of relationship is called basic dependency among attributes.

From the definition of dependency obviously follows that, if $P \rightarrow R$ then $P \rightarrow q$ for every $q \in R$. This kind of dependency is called elementary.

An appropriate use of the above given properties makes us discover all dependencies among attributes in any information table and then consider only the most important attributes in our analysis.

Example 4

Adding attribute c3 to information table of Example 8.1 gives the new coded information table as shown in Table 8.4.

Table 8.4. Coded information table in Example 4.

U	c_1	c_2	c_3	d
x1	0	M	L	0
x2	0	M	L	0
x3	0	M	L	0
x4	0	H	M	1
x5	0	H	M	1
x6	1	L	L	0
x7	1	M	M	0
x8	1	M	M	0
x9	1	M	H	1
x10	1	H	L	1

Let Y = {Y0, Y1} be the following partition of U, Y0 = {x1, x2, x3, x6, x7, x8}, Y1 = {x4, x5, x9, x10} ({d}-elementary sets), and C = {c1, c2, c3}. In this case we have:

$\alpha_C(Y) = \gamma_C(Y) = 1$;

RED_Y (C) = {c2, c3};

CORE (C) = RED_Y (C) ={c2, c3}.

Thus the considered classification could have been obtained taking into account only the two attributes c2 and c3. So these attributes (the core) cannot be eliminated without disturbing the classification.

We observe that {c1, c2, c3}-elementary sets = {c2, c3}-elementary sets = atoms, that is: {x1, x2, x3}, {x4, x5}, {x6}, {x7, x8}, {x9}, {x10}. Thus we have:

set {c1, c2, c3} is dependent in S (IND{c2, c3} = IND{c1, c2, c3});

{c1, c2, c3} $\rightarrow$ {d};

{c2, c3} $\rightarrow$ {c1, c2, c3};

{c2, c3} $\rightarrow$ {d}.

If we indeed consider the set of attributes Q = {c1, c2, c3, d}, we obtain two reducts:
$RED^1(Q) = \{c2,c3\}$, $RED^2(Q) = \{c2, d\}$;
$\{c2, c3\} \cap \{c2, d\} = \{c2\}$ is the core of Q;
$\{c2, c3\} \rightarrow \{c1, d\}$, $\{c2, d\} \rightarrow \{c1, c3\}$ (basic dependencies);
$\{c2, c3\} \rightarrow \{c1\}$, $\{c2, d\} \rightarrow \{c3\}$ (elementary dependencies).

8.2.4 Decision rules

Let objects be independently classified by an expert (partition of U into different classes). In this case condition and decision attributes can be distinguished in the information table S: objects are described by values of condition attributes, while classifications of experts are represented by values of decision attributes (Slowinski, 1992a, 1993). The set of condition attributes is then denoted by C and the set of decision attributes by D ($Q = C \cup D$, $C \cap D = \varnothing$); the data table is called a decision table, that formally is $DT = (U, C \cup D, V, f)$.

We say that a decision table DT is deterministic iff $C \rightarrow D$; otherwise it is non-deterministic. The deterministic DT uniquely describes the decisions to be made when some conditions are satisfied. The non-deterministic DT defines a subset of decisions which could be taken under circumstances determined by conditions, i.e. decisions are not uniquely determined by the conditions.

Let $\mathbf{Y} = \{Y_1, Y_2, ..., Y_k\}$ and $\mathbf{X} = \{X_1, X_2, ..., X_n\}$ be the C-definable and the D-definable classification of U. Any row in the decision table is called a decision rule. Formally, a decision rule is a logical statement (implication: if ...., then ...) relating descriptions of classes in terms of condition attributes and in terms of decision attributes, i.e. $Des_C(Y_i) \Rightarrow Des_D(X_j)$ is called a (C,D)-decision rule in S, where $Des_C(Y_i)$, $Des_D(X_j)$ are unique descriptions of the classes Y_i and X_j.

For each class X_j, j = 1, 2, ..., n, the set of decision rules is $\{r_{ij}\} = \{Des_C(Y_i) \Rightarrow Des_D(X_j) : Y_i \cap X_j \neq \varnothing, i = 1, 2, ..., k\}$. A decision rule r_{ij} is deterministic or exact if $Y_i \subseteq X_j$, i.e. if in the decision table there is no rule with the same condition attribute-values but different decision attribute-values; r_{ij} is non-deterministic or approximate otherwise.

Decision rules, deterministic or non-deterministic, are respectively certain or possible (if condition attribute-values are ... then decision attribute-value is ... or, if condition attribute-values are ... then decision attribute values are ... or ...). DT is deterministic if and only if all its decision rules are deterministic, i.e. if there is no object with the same conditions but different decisions.

A decision rule is reduced if its conditions are based on the reduced set of attributes. Derivation of minimal decision rules from a DT is one of the main tasks of the rough set approach. The set of decision rules for all decision classes X_j (j = 1, 2, ..., n) is called the decision algorithm. Various procedures to obtain different decision algorithms and minimal decision rules have been proposed.

Example 5

Consider again the decision table of Example 8.4. It can easily be interpreted as the following multi-attribute sorting problem, the location of a productive activity with negative environmental externalities. Ten sorting examples have been given by an expert, described by condition attributes C and decision attribute D = {d}. The condition attributes are:
c1: environmental pollution associated with the activity meets a standard associated with a critical damage level (1 = yes, 0 = no, qualitative attribute);
c2: monthly monetary value of production (income) (originally quantitative attribute, then coded in qualitative values by means of norms, see example 1);
c3: opportunities for abatement technology applicable to the specific activity (technology) (qualitative attribute, L = low, M = medium, H = high).

The decision attribute d makes a dichotomic partition of the set of alternatives into accepted (1) and rejected (0). From example 4 it is seen that both accuracy and quality of approximation of this partition by the sets C and RED_Y (C) is equal to one. This means that using the corresponding condition attributes one can perfectly approximate the decision, that results a finite union of elementary sets in S (definable set). Then one can reduce the original decision table without any loss of information considering only the attributes of RED_Y (C) (c2, c3).
The sorting rules generated from this reduced decision table are the following:
rule #1: if "income" = low, then reject
rule #1: if "income" = high, then accept
rule #3: if "income" = medium and "technology" = medium, then reject
rule #4: if "income" = medium and "technology" = low, then reject
rule #5: if "technology" = high, then accept.

All these 5 sorting rules are deterministic and have 7 conditions only. It is interesting to see the sorting rules based on attributes c1 and c2:
rule #1: if "income" = low, then reject
rule #2: if "income" = high, then accept
rule #3: if "pollution" = no and "income" = medium, then reject
rule #4: if "pollution" = yes and "income" = medium, then accept or reject.

The rules #1 and #2 are the same in both sorting algorithms; the rule #4 is non-deterministic: it means that if a given site meets the pollution standard and presents a medium income, it is impossible to make an unequivocal decision unless additional information about technology is known.

We finally observe that, using different norms, the results of rough set analysis can of course change. So, for example, one can obtain the following sorting rules based on attributes c1 and c2, but changing the original norms of attribute c2 (see example 1) into:

Norms of attribute c2: 1300, 1800;

original value of c2	coded value of c2
x < 1300	L
1300 < x < 1800	M
1800 < x	H.

In this case rules #1, #2, #3 remain the same, but rule #4 becomes the following exact one: if "pollution" = yes and "income" = medium, then accept.

8.3 Management of Thresholds in Rough Set Analysis

Rough set theory furnishes a tool whose natural application is devoted to qualitative data. Effectively the rough set approach can deal also with quantitative data but before they must be transformed in qualitative data by means of an adequate codification. This is performed by means of a set of thresholds called norms which discretize the measurement scales by which the quantitative data are expressed. The transformation of quantitative data in qualitative data constitutes one of the most problematic questions in the application of rough set analysis. From the user's viewpoint the introduction of the thresholds could represent a methodological advantage because the discretization of the measurement scale for quantitative attributes should represent the actual user's perception of the analyzed phenomenon that can be represented and analyzed in a form which is really understandable by the user. Nevertheless other exigences are considered in the fixation of the thresholds. For example, very often the considered thresholds are those ones associated to some satisfactory approximation of the considered classes.

The use of thresholds implies some loss of information. Moreover some paradoxical situation can derive from the use of thresholds. As an example, a car is considered expensive if its cost is between $30,000 and $60,000 and is very expensive if it cost is over $60,000. If the cost of car A is $59,500 and the cost of car B is $61,500 then we must classify A among the expensive cars and B among the very expensive cars. Effectively, though, the cost of A and B differs by only $1,000 which is a small, relative difference. Some techniques have been proposed to cope with the problem of the threshold.

(1) The weak indiscernibility (Slowinski, 1992b). Traditionally two objects $x,y \in U$ are indiscernible with regard to a quantitative attribute $q \in Q$ if the values of x and y with respect to q are the same or are in the same predefined subinterval. In order to make the concept of indiscernibility more flexible two indiscernibility relations are introduced: strict indiscernibility and weak indiscernibility. With this aim the intervals generated by the thresholds of attribute q are enlarged from each side so that two consecutive subintervals overlap. Then two objects x and $y \in U$ will be considered as strictly indiscernible with regard to q if their values x_q and y_q belong to at only one

enlarged subinterval which is the same for both values. They will be considered as weakly indiscernible if the value x_q and y_q belong to at least one enlarged subinterval and at least one original subinterval is the same for both values. Lower and upper approximations, dependency and decision rules are then determined on the basis of strict and weak indiscernibility.

(2) The optimal fixation of thresholds (Lenarcik and Piasta, 1992, 1993, 1994). This method maximizes the predictive properties of the information system. The method is based on the idea that the thresholds induce a partition of the attribute value spaces: each element of this partition is obtained by considering an interval of values for each considered attribute. The principal partition criterion, which evaluates the goodness of a given set of thresholds and derived rules, is the expected value of the correctness of classification a priori, i.e. the method minimizes the evaluated misclassification rate. The method starts with a completely discriminant coding, i.e. for each attribute the values of the objects concerned are considered in the increasing or decreasing order and the thresholds are fixed as the mean value of each two subsequent values. The procedure comprises four stages:
(a) the appropriateness of the classification is maximized by removing some intermediate thresholds;
(b) some contiguous partitions of the attributes value space are joined in order to further increase the correctness of a classification;
(c) the most probable value of the decision attribute is attached to each element of a partition;
(d) the number of induced rules is decreased by using an auxiliary criterion.

(3) The similarity relation (Slowinski and Vanderpooten, 1995). This approach extends the indiscernibility relation to a weaker one, the similarity relation. The indiscernibility relation is, in fact, an equivalence relation, i.e. it is a reflexive, symmetric and transitive binary relation while the similarity relation is simply a reflexive relation. Given a pair of objects x and y $\in$ U, x and y are defined similar with respect to an attribute $q \in C$ if $|x_q - y_q| \leq \alpha_q y_q + \beta_q$, where α_q and β_q are pre-defined parameters. If x and if are similar for each $q \in P \subseteq C$ then they are defined P-similar. The similarity relation can be introduced in a more sophisticated way by considering two types of conditions called concordance and discordance which represent respectively positive and negative contributions to the credibility of similarity.

(4) The pairwise comparison table (Greco *et al.*, 1995, 1996). Here the objects are pairs of previous examples instead of single previous examples. With respect to each attribute, both conditional and decisional, the evaluations introduced in a decision table are binary relations defined on the original domain of the correspondent attribute. Practically, when examples x and y are compared with respect to attribute q, the difference between the evaluations of x and y is coded as a binary relation. The rough set analysis is performed using the decision table obtained. In order to classify new pairs of objects, the rules

obtained should be used as follows. If the coded differences with respect the conditional attributes of the new pairs of objects exactly match a decision rule one can comprehensively classify this comparison according to the correspondent value of the decisional attribute of such a rule.

How can the threshold problem now be managed practically? The exposed approaches can give useful support in fixing the thresholds but, with the exception of the optimal fixation, some thresholds or some parameters linked with the thresholds must be given. The following operative procedure for fixing the thresholds of a quantitative attribute can be used:

(a) order the values of the objects in the decision table with respect to the considered attribute from the smaller to the bigger; some graphical representation could be very useful in this stage in order to have a better intuition;
(b) determine some groups of values which are similar, i. e. which are close and such that there are approximately an equal number of different objects in each group;
(c) fix the thresholds in such a manner that each group of similar values are in the same intervals;
(d) process the data using the codification;
(e) try to modify the set of thresholds in order to improve the quality of the approximation.

This operative procedure underlies the selection of threshold values in the rough set based applications of meta-analysis in Part C of the book.

8.4 Rough Sets versus Fuzzy Sets

Some important questions relating to any decision problem are explanation and prescription. Explanation of a decision situation means discovering the most important facts and dependencies in a decision problem involving objects described or evaluated by a set of condition attributes and represented by a table, called information table. If this information can be interpreted as preferential information consisting of sorting examples, constructed or observed in the past by an agent and represented by decision attributes, the aim of prescription is to analyze this decision table in order to obtain a global preference model representing a decision policy for the agent that can be used to support decision makers in new decisions.

Rough set theory has been shown to provide a useful tool to tackle a large class of multi-attribute decision problems. Rough set analysis of a decision table can effectively aid decision makers by providing insights into issues such as the search for optimization of the description of objects of the information table; the construction of reducts of the set of condition attributes C; the computation of the core of attributes C; the evaluation of importance of particular attributes and elimination of redundant ones from the considered decision table; as a consequence of previous analysis; the generation of sorting rules using only the relevant attributes of the reduced decision table in order to support a decision policy and advice a decision maker in new but similar decision situations.

The main approaches to deal with uncertainty in decision making often require preliminary or additional information about data and/or subjective evaluations given by a decision maker. Statistics and the theory of evidence require information about probability and basic probability numbers respectively; cardinal utility theory requires subjective evaluations about probability distributions and certainty equivalence. The fuzzy set approach, that effectively deals with uncertainty has an epistemological foundation based on ambiguity and imprecision of information. It aims to model this vagueness by obtaining from a decision-maker information about the degree of membership or the value of possibility. The rough set approach requires no further information about data or correction of possible inconsistencies in the decision table. So the advantage is a drastic reduction of subjective evaluations and the exploitation in a consistent way of all available information.

Both fuzzy sets and rough sets deal with two different faces of an imperfect knowledge; vagueness and indiscernibility. The fuzzy set approach tries to model the ill-defined boundary of a sub-class of this set in a precise way by introducing the concept of membership function. Rough set theory embodies the idea of indiscernibility between objects in a set, i.e. when an information table is not selective enough to univocally distinguish each object of a set using given information and prevent their precise assignment to a class.

From a mathematical point of view, fuzzy sets are a continuous generalization of set-characteristic functions, allowing to generalise in a useful way the concepts of binary relation and the basic connectors of classic logic. Rough sets are only calculus of suitable partitions of the universe of the objects, approximating with a certain accuracy and quality a particular given classification and using minimal subsets of attributes.

Finally, using a useful example from image processing, fuzzy set theory is concerned with the different levels of grey between black and white (degrees of membership function); rough set theory is related to the size of pixels used to approximate contours of images, that is the granularity of knowledge representation. In other words, the granules, corresponding to equivalence classes of an indiscernibility relation (elementary sets), may cause vague information about a decision situation, and consequently ambiguity to related explanation or prescription.

The issues discussed here are relevant in dealing with partly known and partly understood systems, where terminology, theory and data are still in the process of development like the valuation of services generated by complex systems, such as wetland ecosystems. Integrated policy issues where a balance is to be struck between efficiency, equity, sustainability, and political or social feasibility are bound to be hindered by pervasive and fundamental uncertainties such as discussed here. Finally, the study of longer-term dynamic issues, related to evolutionary system change, network development (transport, materials-product chains, inter-sectoral patterns), technological progress, preference formation, and economic growth, is usually difficult because of uncertainties due to non-observed states and surprises.

8.5 A Comparison of Rough Set and Statistical Approaches to Data Analysis

The first stage of any scientific investigation is the determination of some variables which can explain some observations relative to the analyzed phenomenon. In meta-analysis where, for example, by definition, the analysis is based on a sample of previous research, the considered variables include those ones occurring in every individual study, and some others which can be useful for distinguishing between the studies.

In a second stage of the analysis dependency between the variables is checked. Within the standard method of statistics dependency is measured by means of the correlation index. The consequence of a high degree of dependency of a variable with respect to one or more other variables can be that this variable can be considered redundant for the description of the phenomenon because its informative contribution is still contained in the other variables.

Once determined the set of variables, a theory should be composed of a set of coherent propositions which express some relationships among the values of the relevant variables. These relations are expressed formally, by means of some mathematical equations or disequations called formulas. Often a cause-effect relationship between variables is analyzed. In these cases the values of some variables, called dependent variables, are explained by means of the values of other variables, called independent variables. Among the tools to obtain and/or to validate these formulas, statistics has an important role.

As will be clear from this and the previous chapter, in addition to statistics, rough sets can be a very useful tool in data analysis. From a methodological point of view rough set theory can be referred mainly to the artificial intelligence field (AI). In AI much attention has been devoted to the deductive methods. According to Pawlak, however, with respect to its aims the rough set methodology of data analysis is "rather closer to statistical than deductive methods" (Pawlak, 1991). A comparison between statistics and the rough set approach as a tool for data analysis is, therefore, interesting to examine both overlaps and complementarity or independence of outcomes and interpretations.

From the viewpoint of general philosophy, a comparison between rough set analysis and statistics can be sketched by the following observations. Rough set analysis is based on the concept that knowledge consists of the ability to classify some considered objects. Any classification is based on attributes or categories, i.e. some features of the considered objects. Within rough set analysis, therefore, the objects correspond to the observations in statistics. Furthermore as the observations are characterised by the variable values so the objects are characterised by the attribute values. Therefore the attributes have the same function as the variables in the statistical approach.

Within the rough set context the knowledge can be represented in data tables, called information tables, columns which are labelled by attributes and each row represents a piece of information on the corresponding object. Since also in statistics the data to be processed are generally presented in a tabular form, from this point of view there are not any relevant differences between the two approaches.

Like in statistics, within the rough set approach the data are analyzed by looking for dependencies between attributes. Firstly this dependency is used for obtaining minimal sets of attributes. After this dependency is expressed by inference rules, called decision rules, of the form "if ..., then ...". With this aim often a particular type of information table is considered: the decision table. In the decision table the attributes are divided in condition attributes and decision attributes, in order to explain the latter from the former. With respect to statistics the condition attributes have the same function as the independent variables, while the decision attributes have the same function as the dependent variables. Furthermore the decision rules should be compared with the parameter estimation in a regression analysis. From the user viewpoint, the decision rules have the advantage that they can be expressed in natural language rather than by means of a more technical mathematical formulation.

While with respect to the general stage of an empirical investigation, statistics and the rough set approaches are comparable, they represent different theoretic concepts. With statistical methods the data are generally compared with a pre-existing theory and are used for strict hypothesis testing. For example, according to Stewart and Wallis (1981), econometrics, which can be seen as the application of the statistical methodology to the economic phenomena, is concerned with the confrontation of theory with evidence. Instead, in the rough set approach the analysis of data should not linked to a preceding theory, because in this context developing theories is meant as discovering dependencies between attributes (Pawlak, 1991). From this viewpoint, each set of reduct can be considered as the basis of a possible theory whose propositions are the corresponding "if ..., then ..." rules. While statistics mainly deals with validation, application or rejection of pre-existing, possibly well established, theories, rough set analysis is mainly concerned with some attempts to discover some new theories.

There are some technical differences between the rough set approach and statistics. According to Stefanowski (1992) six main observations can be made.

1. Main objectives. Within statistics, efforts are focused on the determination of some numerical values which represent the phenomenon at hand (average, standard deviation, etc.). A particular model is chosen and the statistical methods allow the examination of the parameters of this model in an optimal way. Very often the model structure or specification is based on the aim to simplify the computation. Therefore linear models are often adopted. Contrary to this, in rough set analysis no particular value is calculated to describe a certain phenomenon. Furthermore, no model is fixed *a priori*. Instead this approach tries to determine cause-effect relationships between the relevant attributes directly from the available data, of course, given the choice of condition and decision attributes.

2. Representation of the data. The traditional statistical methods work rather well with continuous data. There are also some procedures to handle discrete data (e.g. via dummy variables in regression analysis). Such procedures are, however, more complicated. Furthermore, continuous variables in general have a certain degree of imprecision (e.g. measurement approximation). The rough set analysis gives a practical answer to both problems. In fact, it is based on qualitative data which are processed in a very natural and intuitive way. The continuous data are also transformed in qualitative data by means of some techniques for discretization of the measurement scales. This constitutes another advantage of rough set analysis because the different discretization techniques allow to model the different types of imprecision contained in the data of the problem at hand. Furthermore, some specific techniques have been proposed in order to process missing data and fuzzy data within the rough set approach (Slowinski and Stefanowski, 1994).

3. Data requirements. The statistical methods require that formal assumptions should be satisfied: the statistical representativeness of the sample, the multivariate normal distribution of the continuous attributes, the homogeneity of the covariance matrices. Moreover, a sufficient number of observations is required. In rough set analysis none of these requirements must be satisfied *a priori.* On the contrary it is possible to obtain some useful indication also from a small number of examples which have no properties of statistical regularity.

4. Aggregation operators. In the statistical methods there is a very large use of aggregation operators like mean value or covariance matrix. These operators average the original data that therefore are not kept during the analysis in their original form. As opposed to this, in each stage of the rough set analysis all original data are processed. This property may be specially relevant for meta-analysis, where the inputs are the results of other studies and therefore from the beginning they are not original data. In these cases to preserve the data from other average procedures is particular relevant for the significance of the final results.

5. Reduction of the data. Also in statistics techniques are available for reducing the variables by considering their discriminative power. The approach, however, to the reduction of the attributes proposed within the rough set approach is intuitively clearer for the users.

6. Final results. The typical final result of a statistical approach is the estimation of the parameters of a mathematical model. Then the mathematical model can be used for prediction or analysis. This is possible by inserting in the model some forecasted values of the considered independent variables. The advantage of this approach is that there is always a formally precise answer to the considered problem. Within a rough set approach the final result is a set of decision rules in "if ..., then ..." form. By matching these rules with the new objects, an exact classification can be obtained in the cases in which the considered object is assigned to a single class, or an approximate classification in the cases in which the objects cannot be univocally assigned to a single class. It is also possible that, by considering the decision rules, induced by previous examples a new object cannot be assigned to any class because it does not match any rule. This possibility is not necessarily a disadvantage for the rough set approach: but rather an admission that in some situations the available knowledge is not sufficient to give an evaluation for all types of new objects. From this point of view, a statistical model can be criticized because it provides a solution even in cases where there is no previous experience. This solution may in fact entail the averaging of very different situations while the obtained result is not necessarily realistic or relevant. However, with the rough set approach it is always possible to determine the original object or set of objects from which a specific rule or set of rules has been derived. Within rough set analysis, therefore, a critical discussion of the examination is always possible by considering directly the original objects.

Besides technical characteristics, another interesting comparison between statistics and the rough set approach pertains to the role of the user. In the application of the statistical methods the analyst has a very important function. For instance, with respect to econometrics Chow (1983) speaks of "art of formulating a good econometric model" and explains that given the same objectives two econometricians come up with two different models in the same manner that, given a set of requirements in the construction of a building, two architects propose two different designs. Anyway, once a model is fixed, there is a very limited possibility of intervention for the user in the statistical approach. On the contrary, the rough set approach requires a constant participation of the user in all the stages of the analysis.

In the first stage it is important to model the imprecision of the data by choosing among different sets of thresholds. Next, the user can choose among the different reducts the one(s) which better represent(s) the essential features of the phenomenon. Then when some rules are determined to represent the phenomenon, the user can judge about the real significance of the results of rough set analysis by a process of backtracking to the objects. Finally, when the decision rules are matched with other objects in order to make some predictions, it is important to compare the new objects with the original ones from which the considered decision rules have been derived.

In view of the above discussion, one can say that rough set theory is complementary to the more traditional methods encountered in the field of economics like measurement theory, fuzzy sets and statistical methods. It is a valuable tool which can be deployed in meta-analyses to deal with various categories of uncertainty-dominated issues.

PART C

APPLICATION OF META-ANALYSIS TO ENVIRONMENTAL CASE STUDIES

CHAPTER 9 MULTIPLIER EFFECTS IN TOURIST REGIONS

9.1 Introduction

In a way, meta-analysis may be conceived of as quasi-controlled laboratory experimentation with a limited sample of observations drawn under different conditions. This means that meta-analysis has to investigate systematic variations in the estimated outcomes, by analyzing the nature of the explanatory variables, the information base (e.g. geographical scale, time periods and geographical location) and the statistical/econometric techniques used. In this way, the responsiveness of a compound system to various background factors can be explored in a way that makes the results more transferable across varying situations, and even to situations where no directly measurable outcomes do exist. Meta-analysis is essentially more a mode of thinking than an unambiguous analytical procedure, ranging from informal analysis methods (e.g. countings or correlations) to more advanced methods (statistical analysis, rough set analysis).

The focus here is on tourism and in particular with regard to island economies. Tourism has become an important global economic sector (e.g. in Europe it represents some 5 percent of the national product), partly as a result of the rise in incomes, and accompanying rises in leisure time and improvements in transport, and partly as a result of environmental decay in countries of origin (Pearce, 1989). This has led to increasing flows of tourists to historical and cultural cities (e.g. Venice, Paris and Amsterdam) and to sites of a natural beauty (e.g. the Bahamas, Santorini and the Seychelles).

For most tourist destinations, the consequences are twofold (Giaoutzi and Nijkamp, 1993). First, tourism makes up a source of additional local or regional revenues to such an extent that it can become a dominant economic sector (e.g. Crete, Hawaii). Second, the visual, cultural or natural attractiveness of the site may be affected or even destroyed by tourism if it takes place at a large scale (e.g. coastal areas in Spain, Italy or Florida) (Pearce and Turner, 1990). This is the tourism paradox (Coccossis and Nijkamp, 1995). Tourism is a way of escaping unfavourable living conditions in one country in order to enjoy temporarily an attractive quality of life elsewhere, but in doing so a transfer of financial resources takes place to the destination country, while at the same time the tourist flows cause an erosion of the basis upon which tourism rests.

Consequently, a tourist destination is faced with a difficult trade-off between additional income versus a decline in environmental quality. It is, thus, of importance to assess the direct and indirect revenues accruing from tourism in relation to environmental conditions in the areas concerned. In the past, several studies have been carried out to assess the economic consequences of tourism, but only a few meta-analytical explorations of the relationship between tourist demand and environmental damage have been conducted (e.g. Smith and Kaoru, 1990a; Walsh *et al.*, 1989; see also Section 4.5.2).

Most studies on the economics of tourism here have focused on assessing the multiplier effects of tourism in a resort. Many of these studies are concerned with island economies, as islands are often attractive tourist places and as the economic consequences are relatively easy to map out (Pleeter, 1980; Nijkamp and Blaas, 1994).

A central concept in tourism impact analysis is the income multiplier. This is defined as the relative change in regional income consequent upon a given relative change in tourist expenditures in the area concerned. There are different kinds of tourist income multipliers; each of them is corresponding to a specific type of exogenous shock. Examples of such shocks are: expenditures by foreign tourists and expenditures by tourists staying in a hotel. The following is limited to the average tourist income multiplier, which is the ratio of total regional income generated to an average unit of tourist expenditure.

Several studies have been carried out to estimate the tourist multiplier for a given site. A cross-sectional comparison of the various outcomes reveals a range of different values of the tourist multiplier. At the pre-meta-analysis level, one can formulate the objective of explaining differences in multiplier values.

The aim is to produce an explanation of why differences exist in the quantitative results obtained. The information obtained is then used to predict a range of multiplier values for a tourist economy in a situation where we only have insights into background variables. Meta-analytical methods are, therefore, used as a tool for transferability of quantitative outcomes in different situations, since they can avoid designing and undertaking painstaking and laborious field surveys which are otherwise necessary to estimate a regional input-output model. The outcomes of the meta-analyses are used to predict tourism multipliers for the island of Lesvos, in Greece.

9.2 Multiple Multipliers

Regional impact assessment of tourism is usually based on multiplier analysis which aims to estimate all regional consequences of additional (tourist) expenditures. The chain of effects after an initial impulse is often described by means of an input-output model, incorporating all transactions between relevant economic sectors (Fletcher, 1989). The environment can also be included as one of the inputs or outputs related to the production of goods and services. In this way, the system-wide impacts of tourist expenditures in a regional economy can be traced (Armstrong and Taylor, 1993). A more simplified approach to the estimation of regional tourist multipliers has been developed by Archer (1976) who made only a distinction between tourism-aligned and non-tourism-aligned sectors. Even here, however, the calculation of multipliers is a tedious and expensive task. Meta-analysis offers a systematic way of using the information which already exists and can thus reduce costs.

The criterion to include a study in our meta-analysis sample was that it calculated a tourist income multiplier. A sample of 11 relevant and officially published case studies on tourist regions from different sources and covering different years was identified (Table 9.1). A serious problem was that each study provided different background information. The meta-analysis is, therefore, limited to background information common to most of the studies. Where possible, other information has been supplemented by using additional sources. The meta-analysis, given the available information, makes a distinction between the following attributes:

1. Type of documentation (DOC): D_r: research paper, D_j: publication in a journal, D_b: publication in a book
2. Geographic feature of the area (GEO): G_s: single island, G_g: island group, G_r: region, G_c: country
3. Year of collection of data (YEA)
4. Research method (REM): R_{io}: input-output model, R_a: Archer method
5. Estimated values of average tourist income multipliers (TIM)
6. Type of tourist attractiveness (ATR): A_s: sun, A_c: culture, A_n: nature, A_m: mixed
7. Quantitative and qualitative features of the tourist area: Population size (POP), Surface (in km^2) (SUR), Tourist arrivals (TOA), Political autonomy (POA) (yes/no)

The environmental implications in these case studies could not always be measured in a direct sense, but it is assumed they follow indirectly from the number of tourist arrivals, the size of the local population and the size of the area. Only in a few cases could the direct environmental implications be mapped out in more detail.

The main results are set out in Table 9.1. It is noted that the range of variation in the tourist multipliers is significant.

9.3 Statistical Meta-analysis of Tourist Multipliers

The application of meta-analysis to the assessment of tourist income multipliers involves important differences when compared to the conventional medical experimentation model. In the medical model the sample size is often higher, the definition of control groups is unambiguous, the estimation of impact parameters is based on a probability model and the impact assessment boils down to a probability statement in terms of the chance that the effect belongs to a given impact interval. In our case of 11 tourist studies, there is no controlled experiment on samples among households, tourists and other economic sectors in the regional economy. Furthermore, the tourist analysis is carried out on a meso level of aggregation, and finally, in these studies the assumption is made that the measured multiplier (derived either via the input-output approach or the Archer model) is also the real one and not subject to stochasticity (Baaijens, 1996).

The meta-analytical approach to the comparison and assessment of tourist income multipliers has two stages; a linear regression model on the moderator variables of the multipliers and rough set analysis of the background variables with respect to the tourist income multipliers.

Linear regression is a standard technique used in meta-analysis. But its usefulness here is limited because there is only a sample of 11 case studies. What is done involves three stages.

Table 9.1. Concise survey table for meta-analysis for tourist areas.

Study	1 DOC	2 GEO	3 YEA	4 REM	5 TIM	6 ATR	7 POP (x 1000)	8 SUR (x 1000)	9 TOA	10 POA
Bahamas (Archer, 1977)	D_r	G_g	1976	R_a	0.7815	A_s	189.9	11,401	1388.0	y
Bermuda (Archer, 1977)	D_r	G_g	1975	R_a	1.0996	A_s	56.6	107	511,4	n
Singapore (Kahn *et al.*, 1990)	D_j	G_i	1983	R_{io}	0.9393	A_c	2501.0	625	2856.6	y
Turkey (Liu, 1984)	D_j	G_c	1981	R_{io}	1.9809	A_m	45,529.0	779,425	1,460.0	y
Niue (Milne, 1992)	D_j	G_i	1987	R_a	0.35	A_n	2.0	258	1.8	n
Cook Islands (Milne, 1987; Milne, 1992)	D_j	G_g	1984	R_a	0.43	A_s	18.0	236	25.6	n
Kiribati (Milne, 1992)	D_j	G_g	1987	R_a	0.37	A_n	66.0	270	2.0	y
Tonga (Milne, 1992)	D_j	G_g	1987	R_a	0.42	A_s	95.0	699	16.1	y
Vanuatu (Milne, 1992)	D_j	G_g	1987	R_a	0.56	A_m	140.0	12,200	17.5	y
Alonnissos (Pepping & De Bruyn, 1991; Giaoutzi & Nijkamp, 1993)	D_b	G_i	1989	R_{io}	0.489	A_s	1.55	83	20.0	n
Okanagan (Var & Quayson, 1985)	D_j	G_r	1977	R_{io}	0.713	A_n	?	21,813	1,400.0	n

List of variables:
DOC = type of documentation
GEO = geographic feature of the area
YEA = year of collection of data
REM = research method
TIM = estimated values of average tourist income multipliers
ATR = type of tourist attractiveness
POP = population size
SUR = surface (in km^2)
TOA = tourist arrivals
POA = political autonomy

□ a *base model* relates tourist multipliers to geographic characteristics such as surface and population. The area and the population size of the region serve as indicators for the diversity of economic activity in the region. The idea is that a region with a high degree of economic activity is able to produce more of the kinds of goods sought by tourists than a region with a low degree of economic diversity. The latter region has to import relatively more goods, which means smaller indirect effects and a lower multiplier value. The hypothesis is, therefore, that there is a positive relationship between the degree of economic diversity and the multiplier. Because there is no uniform measure for economic diversity, the indicators population size and area are used to test this hypothesis. Regression is used to see if there is a positive relationship between the multiplier and population size, and between the multiplier and the size of the area. In addition, it seems plausible that politically independent areas may have a relatively higher multiplier because of their higher degree of import substitution.

- a *meta model* where the typical meta variables are included in the explanatory analysis - e.g. the source of the study and the analysis method used. In this case, dummy variables may be introduced. A plausible hypothesis is that the input-output model will generate higher multiplier values than the Archer model, because it addresses system-wide economic effects more thoroughly.
- a *tourist model* where the size of the multiplier will be linked to incoming tourist flows. A tourist area which receives a relatively large number of tourists in comparison to the number of inhabitants, is expected to have a relatively low multiplier. The idea is that the tourist sector will crowd out other sectors in case of too many tourists (with respect to labour and capital). As a consequence, the economic diversity in the region will decline, which leads to lower multiplier values. Furthermore, it seems likely that a more equitable distribution of tourism in a country will lead to higher multiplier values (absence of leakage effects). It also seems likely that local property rights and local private involvement in the tourist sector will have a positive influence on the tourist income multiplier, but this is difficult to test.

These hypotheses can only partly and provisionally be tested because of the limited sample.

Base model

In the base model a series of experimental regression analyses on the base variables is carried out. The most important results are contained in Table 9.2. Due to the small sample, only indicative conclusions can be drawn.

It turns out that the tourist multiplier is, in general, positively correlated with the (natural logarithm of the) size of the population. It turns out that two areas have a high value of the residual, *viz.* Turkey and Bermuda. Therefore, some new experiments have been carried out with Turkey and/or Bermuda as a dummy variable. These outcomes show that regions with a larger population size have a relatively larger tourist income multiplier.

Next, also the size of the area has been investigated. The regression results appear to support the hypothesis that regions with a larger surface have a higher tourist multiplier. Since again Turkey appears to have large residuals, a dummy variable has been included for this area. In that case, the results become far less significant, so that a positive relation between area size and the tourist multiplier is not plausible.

Table 9.2. Estimations of base meta-regression models.

var	1	2	3	4	5	6	7	8
const	-0.79* (0.40)	-0.11 (0.41)	-0.87* (0.38)	0.87 (0.44)	-0.095 (0.33)	0.49 (0.31)	-1.29** (0.31)	-0.75 (0.39)
ln (POP)	0.13** (0.034)	0.066 (0.037)	0.14** (0.032)	0.064* (0.024)			0.21** (0.032)	0.15** (0.043)
ln (SUR)					0.11** (0.041)	0.019 (0.044)		
BE			0.48 (0.31)	0.55** (0.16)				
TU		0.92** (0.36)				1.24** (0.41)		0.56 (0.30)
GEO				-1.00** (0.23)				
POA							-0.64** (0.19)	-0.48** (0.19)
R^2	0.66	0.82	0.75	0.94	0.45	0.59	0.87	0.92
n	10	10	10	10	11	11	10	10

Notes: standard deviations are given in brackets.

**	=	the null hypothesis that this coefficient is zero, is rejected at a significance level of 5%
*	=	the null hypothesis that this coefficient is zero, is not rejected at a significance level of 5%, but is rejected at a significance level of 10%
POP	=	population size
SUR	=	surface in km^2
GEO	=	1 if the region is an island or archipelago 0 otherwise
POA	=	1 if the region is an independent nation 0 if the region is not independent
TU	=	1 if the region is Turkey 0 otherwise
BE	=	1 if the region is Bermuda 0 otherwise
n	=	number of observations

Finally, the impact of the degree of political autonomy has also been examined. We come here to the conclusion that no unambiguous impact from the degree of political independence of an area can be found.

Meta model

The results of the basic meta-regression analysis can be found in Table 9.3. The impact of population size in the meta model is significant and positive. Apart from the moderating impact of including Turkey as a dummy, the estimates of multipliers, published in scientific journals, appear to be lower than those published elsewhere; there is apparently, on average, a tendency towards some overestimation in less more informal publication channels.

Further, it is also noteworthy that the type of model used has an impact on the results. The Archer model, for example, tends to yield, on average, lower values of the tourist income multiplier than the input-output model.

Table 9.3. Estimations of meta-regression models with meta variables.

var	9	10	11	12
const	-0.73*	-0.041	-0.41	0.0025
	(0.36)	(0.28)	(0.51)	(0.47)
ln(POP)	0.15**	0.082**	0.12**	0.064
	(0.032)	(0.026)	(0.037)	(0.039)
TU		0.93**		0.84*
		(0.25)		(0.40)
DOC	-0.35	-0.36**		
	(0.20)	(0.12)		
REM			-0.28	-0.12
			(0.23)	(0.20)
R^2	0.76	0.93	0.71	0.83
n	10	10	10	10

Notes:
DOC = 1 if the documentation type is a journal article,
0 otherwise.
REM = 1 if the TIM has been calculated with the model of Archer,
0 if the TIM has been calculated with an input-output model.

Table 9.4. Model estimations with tourist variables.

var	13	14	15	16
const	-1.04**	-0.33	-1.25**	-0.57
	(0.44)	(0.25)	(0.41)	(0.32)
ln(POP)	0.092*	-0.0051	0.16**	0.096**
	(0.046)	(0.029)	(0.031)	(0.027)
TU		1.09**		0.86**
		(0.22)		(0.24)
ln(TOA)	0.066	0.092**		
	(0.052)	(0.025)		
TI			0.043*	0.039**
			(0.021)	(0.013)
R^2	0.64	0.95	0.79	0.93
n	10	10	10	10

Notes:
TOA = number of tourist arrivals in the year of investigation
TI = tourist index: the ratio of the number of arrivals (TOA) to the population size (POP)

Tourist model

An additional approach is to include the tourism-specific variables, in addition to the previous significant variables of population size and Turkey (dummy). The results are contained in Table 9.4. There is some indication that higher tourist arrivals lead to a higher multiplier value, but this is in the context of interactive effects from the dummy variable for Turkey. The impact of the tourist index based on the ratio of incoming tourists to population size, however, is positive.

The statistical meaning of these provisional results is, however, limited. They are merely indicative, but nevertheless interesting, as they generate plausible ideas on the impacts of base, moderator and tourism-specific variables, which may be transferred to other situations.

9.4 Rough Set Analysis of Tourist Multipliers

An alternative to the regression analysis is the application of rough-set analysis. This entails two stages.

- *Information selection.* This means that a distinction has to be made between meta variables (such as source of information, year of publication or statistical model used) and background variables (such as population size, surface of the area, geophysical features, political autonomy, number of tourists, tourist index or type of holiday). This information is usually contained in a survey table as shown in Table 9.1. It contains information (study characteristics) that is known for almost all tourist areas considered.
- *Classification of information.* All objects (in this case, tourist areas) are classified into various categories for each attribute separately. This applies to both categorical and ratio information. In general, some sensitivity analysis on the classification used is meaningful, as a balance has to be found between homogeneity and class size. This classification exercise leads then to a decision table, in which all studies are subdivided into distinct classes for each relevant attribute.

Table 9.5 gives a concise presentation of the results for the 11 case studies examined on the basis of categorical attributes. This table with the condition and decision attributes allows to derive a decision algorithm within a rough set context. The first question is then whether all information in Table 9.5 is necessary for a consistent decision algorithm. In other words: are there reducts in the condition attributes?

It turns out that there is some redundancy in the decision table with 21 different reducts of condition attributes. This relatively large number is, however, plausible, as the number of condition attributes is relatively large compared to the number of objects.

Next, there is the issue of where some condition attributes are more important than others. Table 9.6 provides the answer by showing how often the frequency of various attributes in a reduct.

Table 9.5. Categorical condition and decision attributes from survey table.

	Attributes										
Cases	1	2	3	4	5	6	7	8	9	10	d
1	1	1	3	3	2	1	3	3	1	1	2
2	1	1	2	1	2	2	2	3	1	1	3
3	2	2	5	2	1	1	3	4	2	2	2
4	2	2	6	4	3	1	3	4	2	3	4
5	2	3	1	1	1	2	1	3	1	2	1
6	2	2	2	1	2	2	2	1	1	1	1
7	2	3	2	1	2	1	1	4	1	2	1
8	2	3	2	2	2	1	2	1	1	1	1
9	2	3	3	3	2	1	2	2	1	3	2
10	3	3	1	1	1	2	2	1	2	1	1
11	2	1	7	3	4	2	3	4	2	2	2

Meta variables

1 Type of documentation (DOC)
Classes: 1 research paper
2 article in journal
3 book

2 Year of collection of data (YEA)
Classes: 1 until 1979
2 1980 until 1985
3 from 1986

9 Research method (REM)
Classes: 1 ad-hoc model of Archer
2 input-output model

Geographic variables

3 Population size (POP)
Classes: 1 ≤ 10,000
2 10,000-100,000
3 100,000-200,000
4 200,000-1,000,000
5 1,000,000-5,000,000
6 > 5,000,000
7 no available information

4 Surface (in km^2) (SUR)
Classes: 1 ≤ 500
2 500-10,000
3 10,000-100,000
4 > 100,000

5 Geographic feature of the area (GEO)
Classes: 1 island
2 archipelago
3 country (no island)
4 region (no island)

6 Political autonomy (POA)
Classes: 1 independent
2 not independent

Tourist variables

7 Number of tourist arrivals (TOA)
Classes: 1 ≤ 10,000
2 10,000-700,000
3 > 700,000

8 Share of arrivals from most important country of origin in total number of arrivals ('largest country share' (LCS))
Classes: 1 ≤ 45%
2 45%-75%
3 > 75%
4 no available information

10 Type of tourist attractiveness (ATR)
Classes: 1 "sun, sand and beach holidays"
2 nature and culture holidays
3 "mixed", i.e. both "sun, sand and beach holidays" and nature and culture holidays

Decision variable

d Average tourist income multiplier (TIM)
Classes: 1 ≤ 0.5
2 0.5-1.0
3 1.0-1.5
4 > 1.5

Reducts of condition attributes:
with meta variables: {2,7,10}, {1,7,10}, {2,6,10}, {1,4,10}, {2,4}, {1,5,6,10}, {1,6,9,10}, {1,3}, {2,3}, {1,4,9}, {1,4,8}, {1,4,7}, {1,4,5}, {6,8,9,10}
without meta variables: {3,8}, {7,8,10}, {4,8,10}, {5,6,8}, {5,7,8}, {4,7,8}, {4,5,8}

Table 9.6. Frequency of attributes in reducts.

Attribute	Number (%)
1. Type of documentation	9 (42.9%)
2. Year of collection of data	4 (19.0%)
9. Research method	3 (14.3%)
3. Population size	2 + 1 (14.3%)
4. Surface	6 + 2 (38.1%)
5. Geographic feature of the area	2 + 3 (23.8%)
6. Political autonomy	4 + 1 (23.8%)
7. Number of tourist arrivals	3 + 3 (28.6%)
8. Largest country share	2 + 7 (42.9%)
10. Type of tourist attractiveness	7 + 2 (42.9%)

Note: For other variables than meta variables, the second column table indicates, first the frequency of their occurrence in reducts with meta variables and second, the frequency of their occurrence in reducts without meta variables.

The table shows that 'type of documentation', 'largest country share' and 'type of tourist attractiveness' are the most pronounced attributes, followed by 'surface'. The attributes 'population size' and 'research method' have the lowest scores. With regard to the attribute 'population size' there are, for example, three reducts: (year of collection of data, population size), (type of documentation, population size) and (largest country share, population size). This means that is impossible to design a consistent decision algorithm in which the attribute 'population size' is contained without including any one of the attributes; 'source of documentation', 'year of collection of data', and 'largest country share'. Each of these reducts forms the basis for a consistent decision algorithm, which may be interpreted as a formal logical statement on the size of the tourist multiplier. Not all such formal statements are meaningful theoretical constructs.

9.5 Decision Algorithms

We now focus on decision algorithms which are able to generate statements on the size of the tourist income multiplier in tourist regions outside the sample. Decision algorithms need to be identified that do not include meta variables, since the latter category is only known for actual case studies. Consequently, a subset of reducts from the 21 isolated needs to be selected so that they do not contain meta variables in the decision algorithms. Each of these reducts contains the attribute 'largest country share'. Apparently, by leaving the meta variables out of consideration, the core of the case study attributes is not empty, as the core is then made up by the characteristic 'largest country share'. Unfortunately, the value of this attribute is not known for each case study, so that the predictive value of some of the associated decision rules is limited. As a result, only 2 out of the 7 reducts can be used to predict an unknown multiplier value.

The first is based on the reduct with attributes 'number of tourist arrivals', 'largest country share', and 'type of tourist attractiveness'. This algorithm includes only tourist variables and may be specified as:

Decision algorithm 1
1. $k_7 \leq 10{,}000 \rightarrow TIM \leq 0.5$ (Niue, Kiribati)
2. $10{,}000 < k_7 \leq 700{,}000$ and $k_8 \leq 45\% \rightarrow TIM \leq 0.5$ (Cook-Islands, Tonga, Alonnissos)
3. $k_7 > 700{,}000$ and k_{10} = "sun-sand-sea" $\rightarrow 0.5 < TIM \leq 1.0$ (Bahamas)
4. $k_7 > 700{,}000$ and k_{10} = "nature-culture" $\rightarrow 0.5 < TIM \leq 1.0$ (Singapore, Okanagan)
5. $10{,}000 < k_7 \leq 700{,}000$ and k_{10} = "mixed" $\rightarrow 0.5 < TIM \leq 1.0$ (Vanuatu)
6. $10{,}000 < k_7 \leq 700{,}000$ and $k_8 > 75\% \rightarrow 1.0 < TIM \leq 1.5$ (Bermuda)
7. $k_7 > 700{,}000$ and k_{10} = "mixed" $\rightarrow TIM > 1.5$ (Turkey)

Note: For each decision rule the geographical areas meeting the respective rule are indicated.

In case of a small number of tourists (less than 10,000) the income multiplier is small as well. The case of an intermediate number of tourists (between 10,000 and 700,000) leads to a small value of the multiplier if the largest country share is smaller than 45%, but to a high value of the multiplier (between 1.0 and 1.5) if that share is greater than 75%. Furthermore, in the case of an intermediate number of tourists and a wide spectrum of possible types of tourist attractions, the multiplier falls in the range 0.5 - 1.0. If a large number of tourists is associated with a limited supply of types of tourist attractions in the area, then the multiplier falls between 0.5 and 1.0.

The second decision algorithm is based on the reduct with the characteristics 'surface', 'number of tourist arrivals' and 'largest country share'. The accompanying logical statement explains the size of the tourist income multiplier employing a geographical variable and two tourist variables.

Decision algorithm 2
1. $k_7 \leq 10{,}000 \rightarrow TIM \leq 0.5$ (Niue, Kiribati)
2. $10{,}000 < k_7 \leq 700{,}000$ and $k_8 \leq 45\% \rightarrow TIM \leq 0.5$ (Cook-Islands, Tonga, Alonnissos)
3. $10{,}000 < k_4 \leq 100{,}000 \rightarrow 0.5 < TIM \leq 1.0$ (Okanagan, Vanuatu, Bahamas)
4. $500 < k_4 \leq 10{,}000$ and $k_7 > 700{,}000 \rightarrow 0.5 < TIM \leq 1.0$ (Singapore)
5. $10{,}000 < k_7 \leq 700{,}000$ and $k_8 > 75\% \rightarrow 1.0 < TIM \leq 1.5$ (Bermuda)
6. $k_4 > 100{,}000 \rightarrow TIM > 1.5$ (Turkey)

It turns out that decision rules 1, 2 and 5 from the first decision algorithm are also contained in the second algorithm. We may also conclude that a tourist region with a surface of more than 100,000 km^2 tends to have a tourist multiplier higher than 1.5, whereas an intermediate area (with a surface in the range 10,000 to 100,000 km^2) has a multiplier between 0.5 and 1.0. Finally, a relatively small area (a surface in the range of 500 to 10.000 km^2) with a large tourist influx from the same country has a tourist income multiplier falling between 0.5 and 1.0.

The sensitivity of the results with respect to the base classification in Table 9.1 and 9.5 can be investigated. This has been done in the following way:

- by changing classes of the tourist income multiplier, the particular case of 3 classes (instead of 4 classes) is considered.

- by changing the population size classes. The set of 4 classes has been reduced to 3 (apart from the class 'unknown').
- by changing the classes for the tourist arrivals. The second class has been redefined in order to obtain more uniform classes.
- by changing the classes for 'largest country share'. Again 3 classes have been defined: <50%, >50%, and unknown.

Table 9.7. Adjusted codification of attributes from survey table.

	Attributes										
Cases	1	2	3	4	5	6	7	8	9	10	d
1	1	1	3	3	2	1	3	2	1	1	2
2	1	1	2	1	2	2	2	2	1	1	3
3	2	2	3	2	1	1	3	3	2	2	3
4	2	2	3	4	3	1	3	3	2	3	3
5	2	3	1	1	1	2	1	2	1	2	1
6	2	2	1	1	2	2	2	1	1	1	1
7	2	3	2	1	2	1	1	3	1	2	1
8	2	3	2	2	2	1	1	1	1	1	1
9	2	3	2	3	2	1	1	2	1	3	2
10	3	3	1	1	1	2	2	1	2	1	1
11	2	1	4	3	4	2	2	3	2	2	2

The norms and codification of the following variables have changed with respect to those in Table 9.5:

3 Population size
Classes: 1 ≤ 20,000
2 20,000-150,000
3 > 150,000
4 no information available

7 Number of tourist arrivals
Classes: 1 ≤ 18,000
2 18,000-600,000
3 > 600,000

8 Largest country share
Classes: 1 ≤ 50%
2 > 50%
3 no available information

d Average tourist income multiplier
Classes: 1 ≤ 0.5
2 0.5-0.9
3 > 0.9

Reducts of study characteristics:
with meta variables: {2,7,10}, {2,6,10}, {1,4,10}, {1,6,8,9}, {2,6,8}, {2,4}, {6,8,9,10}, {1,6,9,10}, {1,5,7,10}, {1,5,6,10}, {1,6,7,10}, {1,4,8}, {1,3,4}, {2,3,10}, {1,3,10}, {1,4,9}, {1,4,7}, {1,4,5}, {2,3,8}, {1,3,8}
without meta variables: {5,7,8}, {3,8,10}, {3,7,10}, {3,6,10}, {5,6,8}, {6,7,8}, {3,4,10}, {4,7,8}, {4,5,8}, {3,6,8}, {3,4,8}, {3,4,7}, {3,4,6}, {3,7,8}

A new table may be drawn up with coded information. Table 9.7 represents a consistent set of decision algorithms. Due to the reduction in the number of classes, the number of reducts has increased and there are now 34 reducts. The frequency table of attributes in the reducts is seen in Table 9.8. This table reveals that 'population size' and 'type of vacation' have the highest frequency in forming reducts. From the set of 34 reducts, 20 reducts contain meta variables, which makes these reducts inappropriate for prediction purposes for unavailable case studies. Thus, only 14 reducts are kept for further analysis. It turns out, however, that, from these 14 reducts, 10 reducts contain conditions lacking data, which also renders them unsuitable for predictive purposes. Consequently, only 4 reducts are maintained.

Table 9.8. Frequency table of attributes in reducts for adjusted codification.

Attribute	Number (%)
1. Type of documentation	13 (38.2%)
2. Year of collection of data	6 (17.6%)
3. Research method	4 (11.8%)
9. Population size	5 + 9 (41.2%)
4. Surface	6 + 6 (35.3%)
5. Geographical attribute of the area	3 + 3 (17.6%)
6. Political autonomy	7 + 5 (35.3%)
7. Number of tourist arrivals	4 + 6 (29.4%)
8. Largest country share	6 + 9 (36.6%)
10. Type of tourist attractiveness	10 + 4 (41.2%)

Note: For other variables than meta variables this table first indicates the frequency of their occurrence in reducts with meta variables and second, the frequency of their occurrence in reducts without meta variables.

The following decision algorithms can now be derived. Decision algorithm 3 is based on a reduct which contains 'surface of the area', 'population size' and 'number of tourist arrivals'. Regions with less than 20,000 inhabitants emerge with a tourist income multiplier that is less than 0.5. A medium-sized region (between 20,000 and 150,000 people) with a surface between 500 km^2 and 10,000 km^2 also has a multiplier value of less than 0.5, whereas a large region (more than 150,000 inhabitants) with the same surface has a multiplier value greater than 0.9. Finally, a region with a medium-sized number of inhabitants and a small surface (smaller than 500 km^2), and a small number of tourist arrivals, has a multiplier value less than 0.5. If for the same class of region the number of tourists increases (between 10,000 and 700,000), the multiplier rises above 0.9.

The reduct which forms the basis for decision algorithm 4 contains the attributes 'surface', 'number of tourist arrivals' and 'largest country share'. It turns out that regions with a small surface and a low number of tourists have a small multiplier value, whereas the same types of small regions with a high inflow of tourists (more than 700,000) have a multiplier that exceeds 0.9. Finally, the same type of small region with an intermediate tourist inflow has a multiplier value of under 0.5, if the share of tourists from one single country is less than 50%, while it has a multiplier value greater than 0.9, if this share is higher than 50%.

Decision algorithm 3

1. $k_3 \leq 20{,}000 \rightarrow TIM \leq 0.5$ (Alonnissos, Niue, Cook-Islands)
2. $20{,}000 < k_3 \leq 150{,}000$ and $500 < k_4 \leq 100{,}000 \rightarrow TIM \leq 0.5$ (Tonga)
3. $20{,}000 < k_3 \leq 150{,}000$ and $k_4 \leq 500$ and $k_7 \leq 18{,}000 \rightarrow 0.5 < TIM \leq 1.0$ (Kiribati)
4. $10{,}000 < k_4 \leq 100{,}000 \rightarrow 0.5 < TIM \leq 0.9$ (Okanagan, Bahamas, Vanuatu)
5. $k_4 > 100{,}000 \rightarrow TIM > 0.9$ (Turkey)
6. $k_3 > 150{,}000$ and $500 < k_4 \leq 100{,}000 \rightarrow TIM > 0.9$ (Singapore)
7. $20{,}000 < k_3 \leq 150{,}000$ and $k_4 \leq 500$ and $10{,}000 < k_7 \leq 700{,}000 \rightarrow TIM > 0.9$ (Bermuda)

Decision algorithm 4

1. $k_4 \leq 500$ and $k_7 \leq 18{,}000 \rightarrow TIM \leq 0.5$ (Niue, Kiribati)
2. $500 < k_4 \leq 10{,}000$ and $k_7 \leq 18{,}000 \rightarrow TIM \leq 0.5$ (Tonga)
3. $k_4 \leq 500$ and $18{,}000 < k_7 \leq 600{,}000$ and $k_8 \leq 50\% \rightarrow TIM \leq 0.5$ (Alonnissos)
4. $10{,}000 < k_4 \leq 100{,}000 \rightarrow 0.5 < TIM \leq 0.9$ (Bahamas, Vanuatu, Okanagan)
5. $k_4 > 100{,}000 \rightarrow TIM > 0.9$ (Turkey)
6. $500 < k_4 \leq 10{,}000$ and $k_7 > 600{,}000 \rightarrow TIM > 0.9$ (Singapore)
7. $k_4 \leq 500$ and $10{,}000 < k_7 \leq 600{,}000$ and $k_8 > 50\% \rightarrow TIM > 0.9$ (Bermuda)

Decision algorithm 5

1. $k_3 \leq 20{,}000 \rightarrow TIM \leq 0.5$ (Niue, Cook-Islands, Alonnissos)
2. $20{,}000 < k_3 \leq 150{,}000$ and $500 \leq k_4 \leq 10{,}000 \rightarrow TIM \leq 0.5$ (Tonga)
3. $20{,}000 < k_3 \leq 150{,}000$ and $k_4 \leq 500$ and k_6 = "independent" $\rightarrow TIM \leq 0.5$ (Kiribati)
4. $10{,}000 < k_4 \leq 100{,}000 \rightarrow 0.5 < TIM \leq 0.9$ (Bahamas, Vanuatu, Okanagan)
5. $k_4 > 100{,}000 \rightarrow TIM > 0.9$ (Turkey)
6. $k_3 > 150{,}000$ and $500 < k_4 \leq 100{,}000 \rightarrow TIM > 0.9$ (Singapore)
7. $20{,}000 < k_3 \leq 150{,}000$ and $k_4 \leq 500$ and k_6 = "not independent" $\rightarrow TIM > 0.9$ (Bermuda)

Decision algorithm 6

1. $k_3 \leq 20{,}000 \rightarrow TIM \leq 0.5$ (Niue, Cook-Islands, Alonnissos)
2. $20{,}000 < k_3 \leq 150{,}000$ and $500 < k_4 \leq 10{,}000 \rightarrow TIM \leq 0.5$ (Tonga)
3. $20{,}000 < k_3 \leq 150{,}000$ and $k_4 \leq 500$ and k_{10} = "nature-culture" $\rightarrow TIM \leq 0.5$ (Kiribati)
4. $10{,}000 < k_4 \leq 100{,}000 \rightarrow 0.5 < TIM \leq 0.9$ (Bahamas, Vanuatu, Okanagan)
5. $k_4 > 100{,}000 \rightarrow TIM > 0.9$ (Turkey)
6. $k_3 > 150{,}000$ and $500 < k_4 \leq 100{,}000 \rightarrow TIM > 0.9$ (Singapore)
7. $20{,}000 < k_3 \leq 150{,}000$ and $k_4 \leq 500$ and k_{10} = "sun-sand-sea" $\rightarrow TIM > 0.9$ (Bermuda)

In all cases a large region has a high multiplier value, and a medium-size region an intermediate value. Decision algorithms 5 and 6 are almost identical to decision algorithm 3. They are all based on the reducts 'population size' and 'surface of the area'. The main difference is that algorithm 5 contains 'degree of political autonomy' instead of 'number of tourist arrivals'. These distinct characteristics allow a distinction to be made between Kiribati and Bermuda. Although both belong to the same population and surface class, they differ in terms of 'tourist arrivals', 'political autonomy' and 'type of tourist attractiveness'. As a result, the tourist income multiplier of Bermuda exceeds 0.9, whereas that of Kiribati is less than 0.5. The difference with algorithm 3 is that algorithm 5 contains values of the attribute 'degree of political autonomy' instead of 'the number of tourist arrivals', while algorithm 6 contains values of the attribute 'type of holiday' instead of 'number of tourist arrivals'. These distinct characteristics allow us to make a distinction between Kiribati and Bermuda. Although both belong to the same population area size class, they differ interms of 'tourist arrivals', 'political autonomy' and 'type of tourist attractiveness'. As a result, the tourist income multiplier of Bermuda is higher than 0.9, whereas that of Kiribati is lower than 0.5.

The rough set approach is shown to offer a useful analytical method for classifying tourist regions according to classes of the tourist income multiplier on the basis of various background variables. The results reveal a positive relationship between the value of the tourist income multiplier and the population size, the number of tourists and the tourist market share of the major country of origin.

From a methodological perspective, rough set analysis has the important potential of being able to construct a set of reducts of case study features. Such reducts contain sets of characteristics which discriminate between the various case study results. In this way it is possible, through the elimination of redundancy, to describe the available information using a lower number of attributes.

Finally, an extension of the original set of cases leads to additional information which will affect the composition of the reducts and, hence, the structure of the decision algorithm. This is plausible, as new information should lead to adjusted decision rules; otherwise, the new information is redundant. It is necessary, therefore, to include in a rough set analysis all available information on all available objects or case studies.

9.6 Using Meta-analysis to Predict Unknown Tourist Income Multipliers

The results from both the regression analysis and the rough set analysis can be used to meta-predict an unknown tourist income multiplier for another region, *viz.* the Greek island Lesvos. Relevant background information on Lesvos is given in Table 9.9.

Initially, the results from the regression analysis from Subsection 9.3 are used. Only those specifications of the regression analysis which do not contain meta variables are adopted and the results are contained in Table 9.10. This table does not only show the predictions of the average tourist income multiplier with the help of each of the regression equations, but it also gives the standard errors of these predicted values and their 95% confidence intervals. These intervals show that the results have to be interpreted with great caution. It appears that all regression experiments predict a multiplier value higher than 0.5, but only two experiments predict a value above 0.9. Since the number of tourists arriving to Lesvos is not known exactly, a high and a low estimate for cases where tourist arrivals play an explanatory role is also presented. Passengers arriving in Lesvos can be divided into two groups. The first consists of people who arrive by scheduled flight and by regular ferry boat services. The second group consists of people who come by charter flights. People belonging to this second group are almost certainly tourists. The nationalities, places of origin and motives for travel of those in the first group are unknown. The number of people belonging to the first and the second group is taken as the upper bound for the number of tourists coming to Lesvos. The number of Greeks staying in hotels is known and if we suppose they are all domestic tourists, we may take this number, augmented by the number of people arriving by charter flight, as a lower bound.

Table 9.9. Background information on Lesvos.

Attribute	
Population (in 1991)	86,907
Surface	1630 km^2
"Island"	Lesvos is an island
Political autonomy	Lesvos belongs to Greece
Number of tourists	
- upper estimation	347,545
- lower estimation	47,347+31,511 = 78,858
Share	56.4% of the tourists are Greek
Type of holiday	Both sun and culture holidays
Tourist index	
- lower estimation	0.91
- upper estimation	4.00

Notes:
- upper estimation of the number of tourists = total number of people that arrived in 1991 by boat or airplane on Lesvos
- lower estimation of the number of tourists = number of people that arrived in 1991 by charter flight (foreigners) added with the number of Greeks that in 1990 stayed at least one night in a hotel on Lesvos
- share = quotient of the number of Greeks that stayed in 1990 at least one night in a hotel and the total number of people that stayed in 1990 at least one night in a hotel in Lesvos (100%)
- lower estimation of tourist index = quotient of the lower estimation of the number of tourists and the number of inhabitants
- upper estimation of tourist index = quotient of the upper estimation of the number of tourists and the number of inhabitants

Regression models using the upper estimate of the number of tourists appear to yield higher predictions of tourist income multipliers than regression models using lower estimates of the number of tourists. The differences between the respective predicted values are 0.09, 0.14, 0.14 and 0.12 for the models 13 to 16 in Table 9.4.

Table 9.10. Meta-analytic prediction of tourist income variables for Lesvos using regression analysis.

no.	prediction	dg	interval	no.	prediction	dg	interval
1	0.73 (0.33)	8	[-0.029; 1.49]	13L	0.75 (0.32)	7	[-0.037; 1.50]
2	0.64 (0.25)	7	[-0.042; 1.24]	13H	0.84 (0.33)	7	[0.066; 1.62]
3	0.68 (0.30)	7	[-0.039; 1.40]	14L	0.65 (0.15)	6	[0.28; 1.02]
4	0.58 (0.16)	6	[0.19; 0.98]	14H	0.79 (0.16)	6	[0.40; 1.17]
5	0.74 (0.39)	9	[-0.14; 1.63]	15L	0.62 (0.28)	7	[-0.043; 1.29]
6	0.63 (0.29)	8	[-0.030; 1.29]	15H	0.76 (0.28)	7	[0.10; 1.41]
7	1.11 (0.26)	7	[0.53; 1.69]	16L	0.55 (0.18)	6	[0.12; 0.98]
8	0.96 (0.23)	6	[0.41; 1.52]	16H	0.67 (0.17)	6	[0.25; 1.10]

Notes: standard deviations are given in brackets.
no = number of regression equation (see Tables 9.2 and 9.4)
L = the number of tourists arriving on Lesvos is underestimated
H = the number of tourists arriving on Lesvos is overestimated
dg = number of degrees of freedom of t-statistic

Rough set analysis can be applied to the Lesvos case. This is done by using the six decision algorithms above, excluding all meta variables. This means that the background information on Lesvos contained in Table 9.9 is confronted with the conditions in the decision algorithms. This can lead to three possible outcomes for each of the decision algorithms. First, Lesvos does not obey any of the predecessors in the decision algorithm, so that it is not possible to predict Lesvos' multiplier value on the basis of one of the decision rules. Second, Lesvos obeys the predecessor of exactly one of the decision rules. This outcome is most desirable, since then the decision rule concerned simply gives us one prediction interval. Third, Lesvos obeys two or more predecessors. If in that case the rules concerned prescribe different values of the multiplier, the decision algorithm is inconsistent with respect to Lesvos.

Using the decision algorithms based on the first classification, decision algorithm 1 predicts a multiplier value between 0.5 and 1.0; in that case, Lesvos obeys the value intervals of the decision attributes contained in the fifth decision rule. The second algorithm is inconclusive with respect to Lesvos, as Lesvos does not obey these predecessors. Decision algorithms from 3 to 6 are based on the second classification (Table 9.7). Lesvos obeys a predecessor of a rule that forms a part of algorithms 3, 5 and 6. According to this rule Lesvos has a multiplier value of less than 0.5. Finally, algorithm 4 is inconclusive with respect to Lesvos.

Now the intriguing question emerges whether the use of rough set analysis will allow us to demarcate the feasible area of the tourist multiplier for Lesvos more precisely.

The previous rough set analysis may be helpful by applying the principle of maximum consistency. The interval of multiplier values falling in the range 0.5 to 1 was found by means of decision algorithm 1, based on the first classification of the study characteristics. Given the fact that other island economies inour sample sharing the same characteristics as Lesvos tend to have a higher value of the tourist multiplier than 0.5, it seems plausible to assume that 0.5 is the lower bound for the multiplier of Lesvos.

Then the next question is whether there is also an upper bound. Since the specific features of model 16 in our tourist regression analysis are also the most plausible ones for Lesvos, it seems to be a reasonable inference from our information base that 0.67 is an upper bound of the average tourist income multiplier of Lesvos.

In conclusion, despite the limited number of case studies available, meta-analysis techniques have been able to generate values of response variables, given the existing knowledge base on case studies, including information on site-specific conditions and the particular methodology used.

CHAPTER 10 AIR QUALITY AND PROPERTY VALUE[1]

10.1 Introduction

Since the beginning of the 1970s environmental economists have devoted considerable effort to assessing the cost of environmental damage and the benefits of environmental policy. The economic concept of value underlying these costs and benefits is usually based on neoclassical welfare economics. Direct economic valuation approaches or methods have usually been based on market data. Indirect methods have been based on complementarity or substitution relationships between marketed goods or services and environmental quality, and have led to a variety of empirical methods, the most common of which include travel cost techniques and hedonic property and wage models (see also Chapter 11). For more details see Freeman (1993).

This chapter presents meta-analyses of hedonic property models for estimating the marginal willingness of people to pay for a reduction in the local concentration of certain air pollutants. Many studies on hedonic prices has been undertaken in the past years, but they have not always led to a conclusive answer. In this chapter, based on a large sample of estimated hedonic property models, a meta-analysis will be carried out. The database used was initially developed by V.K. Smith and has been used for a statistical meta-analysis (Smith and Huang, 1995). This allows the skipping of decisions regarding pre-meta-analysis and study selection levels and go immediately to the meta-analysis level. An alternative analysis is adopted, based on rough set theory, to offer more detailed insights and to allow for a comparison of our findings with the Smith and Huang results. The method uses discrete classes to codify values of dependent variables - which are called conditional attributes - and independent variables - which are called decision attributes.

As discussed in depth in Part B of this book, rough set analysis can offer the following:

(a) evaluation of a relevance of particular condition attributes;
(b) construction of a minimal subset of variables ensuring the same quality of description as the whole set, i.e. reducts of the set of attributes;
(c) intersection of those reducts giving a core of attributes that cannot be eliminated without disturbing the quality of description of the set of attributes;
(d) elimination of irrelevant attributes;
(e) generation of "if ..., then ..." rules which synthesize the data analyzed.

[1] The authors wish to thank V. Kerry Smith who has been so kind to allow use of his database for an alternative meta-analysis based on rough sets, as documented in this chapter.

10.2 The Sample for Meta-analysis: Air Quality and Hedonic Property Value Models

Economists have given much attention to empirical studies evaluating air quality using hedonic property prices. The underlying idea goes back to the theory of rents, which states that the equilibrium price of a parcel of land (or of real estate in general) equals the present value of the future flow of net benefits generated (see Freeman (1993), p.368). A variation in property prices may then result from productivity differences, which in turn may be related to environmental variables including air, water or soil pollution factors. In this context, the main questions, as first raised by Ridker (1967), are whether environmental quality changes have systematic effects on property values, whether such effects can be statistically estimated, and how accurately the estimated effects reflect welfare changes as a result of changes in environmental quality.

The hedonic price model allows compensating property price differentials to be interpreted as monetary valuations by property owners, due to differences in local environmental quality, in particular air quality. Usually it is assumed in doing this that property markets operate perfectly. In that case the change in the price of a house resulting from a change in specific characteristics is, *ceteris paribus*, equal to the marginal bid by the buyer and acceptance price of the seller. The estimated coefficient of a particular characteristic or attribute may, therefore, be regarded as the marginal willingness to pay (MWTP) for changes in that characteristic. Different hedonic property studies provide a number of MWTP estimates, which can be compared and summarized by applying meta-analysis techniques.

Economic valuation studies, and hedonic property models in particular, provide a research area which offers relatively large samples for performing meta-analysis. Furthermore, empirical studies have often involved multiple estimations, based on, among others, specific sub-sets of the database, specifications and specific regression techniques, all of which provide potentially interesting observations for a meta-analysis sample.

The suggestion for a relationship between variation in property values and air quality differences was first presented by Ridker (1967), and Ridker and Henning (1967). Since then the literature has expanded considerably. Smith and Huang (1995) have reviewed 50 studies between 1967 and 1988, and identified 37 studies offering hedonic price function estimations including air pollution measures. Table 10.1 lists the studies used in their meta-analysis. The sample will be used in the rough set analysis presented in this chapter.

These studies all deal with urban areas or markets, and in particular large cities. Pooling estimated relationships across studies and model specifications resulted in 167 hedonic model estimations, from which 86 were selected by the authors as suitable for the meta-analysis. In the present chapter, their choice is adhered because it allows for consistency between the input of the statistical and rough set analyses, which in turn permits comparisons of the results and conclusions of these two approaches.

Table 10.1. Summary of the air pollution hedonic model meta-analysis sample.

Study	Year[1]	Number of estimates	Location	Published	Census	MWTP Range (1982-84 US $)
Bender et al. (1980)	1970	3	Chicago	journal	70	159.8 to 234.0
Nelson (1978)	1970	4	Washington	journal	70	0 to 1,522.0
Anderson and Crocker (1971)	1960	4	Washington	journal	60	-4.9 to 169.3
Anderson and Crocker (1971)	1960	4	Kansas City	journal	60	16.4 to 31.6
Anderson and Crocker (1971)	1960	4	St. Louis	journal	60	17.0 to 32.7
Smith (1978)	1971	2	Chicago	journal	-	116.0 to 138.1
Krumm (1980)	1971	1	Chicago	journal	-	29.0
Li and Brown (1980)	1971	3	Boston	journal	-	2.7 to 10.8
Polinsky and Rubinfeld (1977)	1960	4	St. Louis	book	60	35.7 to 37.9
Palmquist (1984)	1977	7[2]	Multiple cities	journal	-	0.4 to 173.7
Brookshire et al. (1979)	1977	1	S. California Air Basin	book	-	0577.2
Palmquist (1982)	1977	20[2]	Multiple cities	unpublished	-	-89.5 to 108.9
Palmquist (1983)	1977	14[2]	Multiple cities	unpublished	-	-76.1 to 98.5
Atkinson and Crocker (1982)	1964	1	Chicago	unpublished	-	366.2
Brookshire et al. (1982)	1977	1	Los Angeles	journal	-	149.2
McDonald (1980)	1970	3	Chicago	journal	70	-239.8 to 159.8
Berry (1976)	1968	1	Chicago	journal	-	-1.38
Jackson (1979)	1970	1	Milwaukee	journal	70	551.4
Appel (1980)	1970	2	New York	Ph.D. dissertation	70	159.0 to 191.3
Soskin (1979)	1970	1	Washington	Ph.D. dissertation	70	73.9
Egan (1973)	1960	2	Hartford	Ph.D. dissertation	60	1612 to 1807.8
Brucato et al. (1990)	1972	2	Los Angeles	journal	-	140.7 to 190.6
Brucato et al. (1990)	1978	1	San Francisco	journal	-	500.2

Notes:
References are listed in Smith and Huang (1995, Appendix).
1. For those studies using census tract data, the year refers to the census; for other studies it refers to the year of the housing prices or sales used.
2. The number of estimates equals the number of cities studied.
Source: Smith and Huang (1995) and personal communication with V.K. Smith.

The MWPT, whose ranges are shown in Table 10.1, were measured as the change in the asset value of the property. When rent was used in the primary hedonic price relationships, the estimate was converted into a measure of housing price change. The median MWPT for the sample is $ 22.40, but, as Table 10.1 shows, the range of estimated values is quite wide. Negative values indicate estimates that are not in agreement with *a priori* expectations. It should also be noted that the MWTP is stated in terms of reducing total suspended particulates (TSP), where the TSP measure is adjusted to be in comparable units across studies, using general type of information sources (see further Smith and Huang, 1995).

10.3 Main Conclusions of the Smith and Huang Statistical Meta-analysis

The variables in the Smith Huang meta-analysis are taken from two sources. Variables used in each specific study describing the characteristics of the estimation of hedonic housing price models were included in the meta-analysis sample. Among these are those mentioned in Table 10.1. In addition, information obtained from general sources was matched to this. This includes the geometric mean of the second highest concentrations of TSP by city, the per capita income for each city and the vacancy rate, all for the year closest to the data sample in Table 10.1.

The variables in the meta-analysis are divided into four categories. First, there are the characteristics of households, which includes the TSP measure, the real income in constant dollars (to deal with distributional impacts on the marginal valuation) and the vacancy rate (to correct for a deflating impact on the marginal valuation). The second category includes characteristics of model estimation, covering the number of neighbourhood and air pollution variables, and dummy variables reflecting whether the actual price was used, the model was linear or nonlinear, a semi-log specification was used, a log-log specification was used, and whether an OLS estimator was used. The third class is composed of characteristics of the data such as whether census 1960 and census 1970 data were used, and the year of the data. Finally, characteristics of the meta-sample are represented by referring to the questions whether a study was unpublished (including Ph.D. theses), and whether the Mills ratio for dealing with publication selection effects was used. The meta-analysis data base has a number of statistical properties. First, the MWTP variable may be expected to show heteroscedasticity because the MWTP measures have been estimated with varying levels of precision and many of them are based on nonlinear specifications. Second, the observations in the sample are not independent, because estimation of different specifications for a given study (city) may be based on the same data. In order to take these qualifications into account, two estimation approaches are adopted. The first is to use a minimum absolute deviation (MAD) estimator describing the relationship between the conditional median of the MWTP estimates and real per capita income, the conditions of local housing markets, air pollution, and the features of each original hedonic model estimation such as type of data, model specification features, publication status and the estimation procedure. This approach can address the feature of outlying observations, which make the assumption of normally distributed errors, necessary for OLS estimation, doubtful. The second approach is an OLS estimation with a Huber consistent covariance matrix for the parameter estimates, treating each study's MWTP estimates as identical to a set of sampling clusters. This is useful, as it recognizes the potential correlation between observations, and can deal with heteroscedasticity. It should be noted that the approaches are to some extent complementary and that both can only take part of the peculiarities of the sample into account. Conclusions should, therefore, be drawn on results of the two methods applied to the sample. These results are summarized in Table 10.2, but, in addition, Smith and Huang also presented a meta-analytical specification with only air pollution and real income considered. This is however, incompletely reported (personal communication with V.K. Smith) and has only few significant variables.

Table 10.2. Meta-analysis of MWTP for TSP, using OLS and MAD.

Independent variables	OLS	MAD
Intercept	-4739.02***	-2536.90***
A. Characteristics of households		
1. TSP measure	-1.44**	-0.311**
2. Real income	0.02	0.002
3. Vacancy rate	-73.01	-20.22**
B. Characteristics of model estimation		
4. Number of neighbourhood variables	199.48***	156.05***
5. Number of air pollution variables	-24.38	-13.45**
6. Actual price	3.15	95.92**
7. Linear model	936.66***	690.21***
8. Semi-log	832.17***	681.03***
9. Log-linear	278.65**	244.14***
10. OLS estimator	163.70*	-0.21
C. Characteristics of the data		
11. Census 1960	1470.44***	915.79***
12. Census 1970	456.27***	304.45***
13. Year of the data	51.96**	23.10***
D. Characteristics of the meta-sample		
14. Unpublished study	247.67*	230.12***
15. Inverse Mills ratio	205.22	205.19***
R^2	0.688	0.337

Notes:
1. Number of observations = 86.
2. Significance levels are: * -> p=0.10; ** -> p=0.05; *** -> p=0.01.

Source: Smith and Huang (1995).

The results show that, although the OLS and MAD estimates generally offer similar signs and magnitudes, the significance levels of estimated coefficients are quite different for some variables, notably variables 5, 6 and 15 - i.e. the number of neighbourhood variables in the original studies, the actual price and the inverse Mills ratio. Very significant in both estimations are variables reflecting the number of neighbourhood variables, the dummies for (non)linearity and the semi-log specification, and the variables relating to the year of the data. The latter is explained by the more recent studies using improved model specifications.

Another finding is that smaller MWTP estimates are found for published studies, which may be consistent with the well-known phenomenon of publication bias often mentioned in the meta-analytical literature. In this case, higher estimates may cause pre-selection or dropping of sub-samples and specifications. This selection effect is also caught by the inverse Mills variable, which is derived from a probit sample selection model to see whether sufficient information has been included in the meta-analysis sample. The MAD estimation indicates that this is not the case, which implies that more information, extending the meta-analysis sample, should be collected. This can be based, for instance, on the hedonic price estimations that were dropped from the initial sample of 50 studies, or on newly performed studies.

10.4 Results of the Rough Set Analysis

In this section a meta-analysis of the sample based on rough set theory is performed. Initially, the outcomes are classified for all variables by defining proper thresholds. Variable numbers referred to are in Table 10.1. The number of thresholds is equal to the number of classes less one. For instance, a single threshold t for a variable x means that for $x \leq t$ all values are in a separate class, and for $x > t$ they are in another class. The choice of thresholds for all variables is somewhat subjective, but may be based on their distribution of values in the sample over the total range of possible outcomes, for each particular variable (for more details, see Section 8.3). Such a procedure has for this set of variables led to the choices of thresholds shown in Table 10.3.

Applying these thresholds, or norms, to the meta-analysis sample produces a set of coded variables. Based on a rough set analysis of the sample of 86 cases, four main sets of indicators and outputs can be calculated.

Table 10.3. Thresholds of the variables in the rough set analysis.

Variables	*Thresholds*
Dependent variable: MWTP estimate	0 / 50 / 200 / 1000
Independent variables:	
1. TSP measure	130
2. Real income	15 / 18
3. Vacancy rate	1.2 / 1.45 / 3
4. Number of neighbourhood variables	0.5 / 1.5 / 2.5 / 3.5
5. Number of air pollution variables	1.5 / 2.5 / 3.5 / 4.5
6. Actual price	0.5
7. Linear model	0.5
8. Semi-log	0.5
9. Log-linear	0.5
10. OLS estimator	0.5
11. Census 1960	0.5
12. Census 1970	0.5
13. Year of the data	65 / 75
14. Unpublished study	0.5
15. Inverse Mills ratio	0.3 / 0.7

(1) The *reducts*, i.e. all combinations of independent variables (or, conditional attributes) which can completely determine the variation in the dependent variable (or, decisional attribute) without needing other explanatory variables.

The reducts for the present sample are given in Table 10.4. These may be considered as competitive theories whose factors influence the MWTP value. With each theory or reduct, 9 or 10 factors are required to completely explain the variation in the estimated MWTP values among the various case studies and estimated relationships per case study. The additional variables to the core (see point (2)) to construct reducts, as indicated in Table 10.4, are {4,9,12}, {4,7,8}, {6,8,9,13}, {6,7,8,13}, {4,8,9} and {4,7,9}. This implies that in terms of theorizing based on the reducts, substitution between variables is possible. In this case, linear model, semi-log and census 1970 dummy variables are clearly overlapping; linear model and log-linear dummy variables are overlapping; and semi-log and log-linear dummy variables are overlapping. These are all dummy variables, and most of them are related to the model specification of the original studies. One step further is to compare groups of variables that can substitute for one another, such as the sets of variables {4,12}, {6,8,13}, {4,8} and {4,7}.

Table 10.4. A summary of the results of the rough set analysis.

Reducts	Core	Number of rules (LERS algorithm)	Number of rules (Roughdas algorithm)
{1,3,4,5,9,10,12,14,15} {1,3,4,5,7,8,10,14,15} {1,3,5,6,8,9,10,13,14,15} {1,3,5,6,7,8,10,13,14,15} {1,3,4,5,8,9,10,14,15} {1,3,4,5,7,9,10,14,15}	{1,3,5,10,14,15}	31 rules total Exact • class 1 = 5 • class 2 = 7 • class 3 = 7 • class 4 = 5 • class 5 = 1 Approximate = 6 Rule strength • max = 14 • min = 1 • average = 2.806 Rules length • max = 5 • min = 1 • average = 2.871	38 rules total Exact • class 1 = 6 • class 2 = 10 • class 3 = 10 • class 4 = 5 • class 5 = 1 Approximate = 6 Rule strength • max = 12 • min = 1 • average = 2.263 Rules length • max = 18 • min = 2 • average = 7.500

(2) The *core*, i.e. the set of variables that are in all reducts as discussed under (1), or that are part of all theories. In our case, the core is equal to {1,3,5,10,14,15}. In other words, the TSP measure, the vacancy rate, the number of air pollution variables, the OLS estimator, the unpublished study and the inverse Mills ratio are essential factors in explaining the MWTP estimate variation.

(3) *Rules*, i.e. exact or approximate relationships between explanatory variables and dependent variables. These may be considered as "if ...., then ..."

statements. Tables 10.4 and 10.6 provide information on this. The support of rules by cases is also a useful indicator. If a rule is supported by more objects, then it is more important, for instance, in summarizing the different single case study results. In the present case there are over 30 rules, dependent on the algorithm used, LERS (Grzymala-Busse, 1991 and 1992) or Roughdas (Slowinski and Stefanowski, 1992). Furthermore, these may include approximate rules that have considerable support from cases or objects. It is important, however, also to check whether the supporting objects of a rule belong to a single or to separate studies. In the example of Table 10.6 it is a mix of the two. The target support seems to be for classes 2 and 3, i.e. MWTP estimates in between 0 and 200.

(4) The *lower and upper approximation*, and derived accuracy of relationships for each value class of the decisional variable. From this can also be derived a accuracy and quality of classification, which is related to the ratio of exact and approximate rules. In this case these indicators are less than 1 which is the maximum value, as seen in Table 10.5.

Table 10.5. Accuracy and quality of the classification of the decisional variable (number of objects = 43).

Class of decisional/dependent variable	Accuracy	Lower approximation	Upper approximation
1	0.2941	10	34
2	0.5385	28	52
3	0.5833	14	24
4	1	5	5
5	1	2	2
Accuracy of classification	0.5043		
Quality of classification	0.6860		

Note: The accuracy for each class is the lower divided by the upper approximation.

Table 10.6. Two most supported exact rules generated by the Roughdas algorithm.

Value/class of independent variables	Implied class dependent variable	Number of cases supportive of rule
var1=2, var3=2, var14=1	2	14
var5=2, var8=1, var11=1, var14=1	2	6

Finally, a *sensitivity analysis* can also be performed, notably by changing thresholds or classes of the dependent and core variables. Here we will only focus on a comparison of the statistical with the rough-set analysis results. Chapter 11 devotes more attention to sensitivity analysis and alternative meta-model analyses.

10.5 A Comparison of Methods

Table 10.7 summarises a comparison of the two approaches contrasting the significant variables in the statistical results with the core variables plus other variables in the reducts from the rough set analysis.

Table 10.7. Comparison of different methods applied to meta-analysis of hedonic housing price studies.

Analytical results	OLS	MAD	Rough sets
A. Significant variables			
- p=0.01	4,7,8,11,12	4,7,8,9, 11-15	
- additional at p=0.05	1,9,13	1,3,5,6	
- additional at p=0.1	10,14	-	
B. Core variables			1,3,5,10,14,15
C. Additional reduct variables			4,6,7,8,9,12,13
D. Variables not in A,B or C	2,3,5,6,15	2,10	2,11

Two approaches can be taken to evaluating the information contained in Table 10.7. First, one can compare the sets of significant and core variables. Second, since statistical models can be regarded as complete theories, one can also compare non-significant variables with non-reduct variables. The resultant cross-comparisons yield a number of features. Variable 2 seems unimportant from any perspective, so that the distributional effect on the MWTP can be neglected. Variables 4,7,8 and 11 are most significant (p=0.01) in both statistical results (OLS and MAD). The first three of these are also reduct variables, but none is a core variable. Furthermore, variable 11 (census 1960) is also irrelevant according to the rough set analysis' results. The situation is different when the 95% significance level is adopted. The overlap between OLS and the rough set analysis' results is only variable 1, i.e. the TSP measure. The MAD and rough set analysis results show more overlap i.e. between the TSP measure, the vacancy rate, the number of air pollutants, the unpublished study and the inverse Mills ratio. These variables are thus important according to both methods.

10.6 Conclusions

This chapter has explored the implications of employing rough set analysis, as opposed to regression type models, in conducting meta-analysis. Links between air quality and property values have formed the basis of this work.

The results from the rough set analysis of the application of meta-analysis to hedonic model studies differ somewhat from the results of the statistical analysis. Comparing reduct variables with significant variables in general, leads to the conclusion that the variables TSP measure, number of neighbourhood variables, linear model, semi-log, log-linear, and census 1970 are essential according to all three methods used (OLS, MAD and rough set theory). This implies that model characteristics are especially relevant in explaining differences among studies, whereas characteristics of the data are less essential. Characteristics of the meta-sample, however, appear to have some relevance. With regard to the latter, however, the OLS and MAD results conflict, since OLS indicates that the unpublished study variable is relevant, whereas according to MAD the inverse Mills ratio is relevant. In addition, the rough set analysis shows that both are part of the set of core variables. In other words, the meta-sample characteristics seem relevant after all. To help draw conclusions on the merits of the three different methods, one can also consider which variables are not essential in explaining differences among studies. All three methods support the variable real income. Other variables, supported by only one of the three methods, are vacancy rate, number of air pollution variables, actual price, OLS estimator, census 1960, and inverse Mills ratio. This may offer some guidance for choosing variables and model types when performing additional studies.

CHAPTER 11 WORKING CONDITIONS IN INDUSTRIAL SECTORS: VALUATIONS OF LIFE

11.1 Introduction

As part of the process in deciding about public policy and private management to reduce danger at work, most previous studies have adopted compensating values. This assumes that society or individuals implicitly or explicitly use a value of life equal to the amount of money they are willing to pay to reduce the risk of accidents or mortality. Here, labour market-based evidence on values of life is examined by applying two meta-analysis techniques.

Estimated marginal values of safety and derived values of life from labour market studies show variation. Several authors have already provided literary overviews comparing studies and trying to explain differences (Freeman, 1993; Viscusi, 1993; Johansson, 1995). This study applies two formal methods of meta-analysis - a statistical meta-analysis and an analysis based on rough set theory. These are used to aggregate various individual case-study results and to reduce the influence of case-study specific factors on the estimation of the value of life. The application is possible because the sample consists of studies which are reasonably comparable in terms of data and methods used.

The statistical meta-analysis is based on ordinary least squares and generates average-type output. The alternative analysis is based on rough set theory and offers a new approach to meta-analysis. It uses discrete classes to classify values of variables and generates unique rules, core variables, quality indicators and upper and lower bounds. The application of these approaches generates two sets of outcomes. The interpretations are partly overlapping and partly complementary.

11.2 The Meta-analysis Sample

In the past considerable attention has been devoted to empirical studies of hedonic wage functions in terms of risk associated with working conditions. The hedonic wage model allows one to examine compensating wage differentials as monetary valuations by workers of differences in working conditions. These differences may relate to regional environmental amenities, or to conditions that relate to the risk of injury, illness or death. Since both demand for and supply of labour are affected by as well as influence working conditions, the hedonic wage model represents an equilibrium situation in the labour market. The derivative of the hedonic wage function with respect to a job characteristic can be interpreted as its marginal price. The equilibrium assumption in combination with assumptions regarding utility maximization by workers and profit maximization by employers means that the estimated marginal price equals the workers' marginal willingness to pay for the associated characteristic, and the employers' marginal value of that characteristic. For more details about assumptions see Freeman (1993), Viscusi (1993) and Johansson (1995).

The marginal implicit price of the risk attribute may be regarded as a risk premium. In order to estimate the hedonic wage function, data on wages and job and worker attributes are used. The valuation of risk reduction in working conditions by workers depends on their perceptions, or awareness, of differences in risks across jobs. Estimation of compensating wage differentials for risk may be biased if perceptions of job risks are inaccurate. A problem in performing studies, and consequently in comparing different studies, is the use made of objective measures of job risk, which do not necessarily match workers' perceptions. In fact, various studies comparing objective and subjective risks indicate that workers often overstate risks, although the correlation between the two is positive. Own risks, however, are again often underestimated. Freeman (1993) notes that what matters in relation to the hedonic wage model estimation is how accurate worker's perceptions are of the differences in, instead of absolute levels of, risks among different jobs, but that there seems to be no evidence on this point.

Table 11.1. Some characteristics of individual studies.

Study	year	# used estimates	Range of estimates (current mill. US$)	# sub-samples
Thaler and Rosen (1975)	1976	8	0.14-0.62	813-907
Smith (1983)	1976	NA	-	-
Viscusi (1988)	1978	NA	-	-
Olson (1981)	1981	2	3.20-3.40	5993
Viscusi (1988)	1981	NA	-	-
Marin and Pasacharopoulos (1982)	1982	16	0.39-15.63	688-5509
Smith (1983)	1983	1	2.36	7094-16199
Arnould and Nichols (1983)	1983	2	0.23-0.29	1811
Dillingham (1985)	1985	2	1.60-2.80	514
Moore and Viscusi (1988)	1988	8	1.93-6.00	1349
Gegax *et al.* (1991)	1991	4	0.81-11.84	178-737

Note: NA = not available (see text for explanation).

Table 11.1 shows the sample of studies that are considered in the meta-analysis. All these studies considered measures of risk of death only, i.e. excluding injury. The selection of these studies is based on secondary sources, such as Freeman (1993). The criterion for selection was similarity of the problem investigated and the method used. The sample constructed for the meta-analysis is the result of a pooling of estimated hedonic wage functions across different case studies, model specifications, data-subsets, regions, risk measures, and dependent variable measures. These elements are shown in the table. The size of the resulting sample of past studies is 43, which is a mix of different case studies and different estimated relationships within specific studies. This excludes entire studies as well as estimated relationships for given case studies for which no complete information was available, because these studies were not published in journals or books. From the table one sees that studies differ with regard to the year of publication and the number of estimates generated on the basis of different model specifications or sub-samples. The time to which the data refer is approximated by the year of publication. This assumption is adopted, since the data were in most cases a combination of separate data on risk, labour and economic variables, not necessarily referring to the same year or period, and, therefore, difficult to capture in a single temporal variable.

11.3 Results of the Statistical Meta-analysis

Various statistical analyses using ordinary least squares were performed. Different meta-analytical estimates were obtained by pooling different estimates from various studies, some involving multiple estimated models in the context of particular reported studies. The dependent variable is the value of life which is directly related to the risk coefficient in the estimations, so the latter was excluded from the analysis. The independent variables included are shown in Table 11.2. These embrace mainly model characteristics.

Table 11.2. List of independent variables in the meta-analysis.

1	=	sample size;
2	=	number of variables in model;
3	=	number of risk variables in model;
4	=	dummy for linear (1) or non-linear (0) equation;
5	=	dummy for dependent variable logarithm (1) or not (0);
6	=	number of interaction terms;
7	=	estimated constant;
8	=	t-statistic risk effect;
9	=	estimated value of life;
10	=	estimated R^2;
11	=	indicator for database (unique for each overall study);
12	=	indicator for risk variable (11 types);
13	=	dummy for whether whole sample (1) or subset (0) was used.

In order to test the influence of these variables on the value of life, the following approach was followed:

(i) Analysis of the total 43 cases of the meta-analysis sample. The average value of life estimate in this case is equal to US$ 3.59 million. This means that all estimation models from singular or different studies are regarded to be mutually comparable. Since several of these are based on subsets associated with particular sectors, each of which is associated with a particular level of working condition mortality risk, and as some estimates of the crucial risk variable were not significant, the value of life estimate may not be considered very reliable. The following two additional options are, therefore, also examined.

(ii) A subset of the sample of cases used in option (i), restricted to those estimated relationships that were based on a complete sample in the respective individual studies. In this case, it was hoped that estimated relationships from different studies which were maintained would allow for mutual comparison. As a result, 28 observations remained, with an average value of life estimate as equal to US$ 2.34 million.

(iii) A subset of the total sample, embracing estimated relationships that included the whole sample of respective individual studies and satisfied a minimum level (1.8) of the estimated t-statistic of the risk-effect. This means that only good quality estimates of the risk variable parameter were allowed for in the meta-analysis sample, since this directly affects the quality of the value of life estimate. This case has 13 observations, and an average value of life estimate equal to US$ 3,86 million. It should be noted that the supposedly most reliable average estimate of the value of life is also the highest.

For each of the options (i) to (iii) correlations of all variables are calculated, in order to decide about the combinations of independent variables which show a minimum of correlation among each other. Regressions are conducted on each set of independent variables that exclude combinations of variables with high mutual correlations (> 0.5) according to Table 11.3. The resulting regression estimates for each option are set out in Table 11.4. Since it would take too much space to discuss all these regressions, only some important conclusions are listed in the following.

Table 11.3. High correlation between variables for each option (correlation > 0.5).

variable	option (i) 43 observations	option (ii) 28 observations	option (iii) 13 observations
1	3	3	8
2	4, 11, 12	-	-
3	-	1	-
4	2	-	-
5	7	-	-
6	-	-	-
7	5, 11, 12, 13	11, 12	11, 12
8	-	-	-
9	-	-	-
10	-	-	-
11	2, 12	12	7, 12
12	2, 7, 11	11	7, 11
13	7	not included*	not included*

Note: * Variable 13 takes the value 1 for the entire sample under options (ii) and (iii), as only cases are selected which satisfy this value.

Table 11.4. Estimated relationships for each option, based on Table 11.3.

Estimated relationships	option (i)	option (ii)	option (iii)
combinations of independent variables	a: 1-13 b: 1-10, 13 c: 1,2,5,6,8,10,13 d: 2,3,5,6,7,8,10 e: 2,3,6,7,8,10 f: 1,2,6,7,8,10 g: 1,4,6,7,8,10 h: 1,4,6,7,8,10,11	a: 1,2,4,5,6,7,8,10 b: 2-8,10 c: 1,2,4,5,6,8,10,11 d: 1,2,4,5,6,8,10,12 e: 2-6,8,10,11 f: 2-6,8,10,12	a: 1-7,10 b: 2-7,8,10 c: 1-6,10,11 d: 2-6,8,10,11 e: 1-6,10,12 f: 2-6,8,10,12

Option (i):

- adjusted R^2 ranges from 0.28, regression (a) to 0.59, regression (b);
- significant variables occur in some regressions (t-statistic >2): sample size, R^2, number of variables in model, t-statistic, constant, database, linear;
- regression (b) has most significant variables (#=6): sample size, R^2, number of variables in model, t-statistic, constant, database;
- regressions (a) and (f) have the least significant variables.

Option (ii):

- adjusted R^2 ranges from 0.82 to 0.89;
- significant variables occur in some regression (b) (t-statistic >2): sample size, R^2, number of variables in model, t-statistic, constant, database, indicator for risk variable.

Option (iii):

- adjusted R^2 ranges from 0.92 to 0.98;
- significant variables occur in some regression (b) (t-statistic >2): sample size, R^2, number of variables in model, number of risk variables in model, number of interaction terms in model, t-statistic, constant, database indicator, indicator for risk variable, dummy for linearity and logarithmic dependent variable.

These regression results show that many variables can explain the variation in the value of life estimates.

11.4 Results of the Rough Set Analysis

In this section a rough set analysis of the data given in Table 11.1 is performed. This is done by initially classifying the outcomes for all variables by defining thresholds. The number of thresholds is equal to the number of classes less one. A threshold t for a variable x means that for $x \leq t$ all values are in a separate class, and for $x > t$ they are in another class. The choice of thresholds is somewhat subjective, but may be based on the distribution of values or ranges of values over the total range of possible outcomes, for each particular variable. Such a procedure in the present case gave rise to the sets of thresholds in Table 11.5.

Table 11.5. Thresholds of the variables in the analysis.

Variables	*Thresholds*
1= sample size	600/1000/1500/5000/10000
2= number of variables in model	15/30
3= number of risk variables in model	2
4= dummy for (non)linearity	0.5
5= dummy for logarithm	0.5
6= number of interaction terms	0.5/1.5
7= estimated constant	1
8= t-statistic risk effect	1.8/3
9= estimated value of life	1/4/8 (million)
10=estimated R^2	0.5/0.7
11=indicator for database	1.5/2.5/3.5/4.5/5.5/6.5/7.5
12=indicator for risk var.	2.5/3.5/4.5/5.5/6.5/7.5/8.5/9.5/10.5
13=dummy subset	0.5

Applying these thresholds to the set of data produces a set of coded variables. Based on a rough set analysis of the sample of 43 cases as indicated in the previous section, several sets of indicators and outcomes can be calculated.

(1) The reducts, i.e. all combinations of explanatory or independent variables which can completely determine the variation in the dependent variable, or decisional attributes, without needing other explanatory variables. For the present sample, the reducts are: {1,8,12,13}, {1,8,11,13}, {1,6,8,13}, {1,3,7,8,13} and {1,2,8,13}. These may be considered as competitive theories of factors influencing the value of life. In each theory or reduct, four or five factors are sufficient to explain completely the variation in the estimated values of life among the various case studies and estimated relationships per case study.

(2) The core, i.e. the set of variables that are in all reducts, or that are part of all theories. In the present case it is equal to {1,8,13}. In other words, the sample size, the t-statistic of the risk effect and the subset of the database are essential, but not a complete set of factors in explaining the variation in value of life estimates. The inclusion of the t-statistic among the essential factors can be explained by noting that the value of life estimate is derived from the risk effect, so that its value may depend on the quality of the risk effect estimate - the associated t-static.

(3) Rules, i.e. exact and approximate relationships between explanatory variables and dependent/decisional variables. In this case the number of exact rules is equal to 18. These may be considered as "if ...., then ..." statements.

(4) The lower and upper approximation, and derived accuracy of relationships for each value class of the decisional variable. From this can also be derived a accuracy and quality of classification (i.e. choice of thresholds), which is

related to the ratio of exact and approximate rules. In this case these indicators are less than 1, as indicated in Table 11.6.

Table 11.6. Accuracy and quality of the classification of the decisional variable (number of objects = 43).

Class of decisional/ dependent variable	Accuracy	Lower approximation	Upper approximation
1	1	13	13
2	0.55	11	20
3	0.5	6	12
4	0.5714	4	7
Accuracy of classification	0.6538		
Quality of classification	0.7907		

The options (ii) and (iii) lead to different outcomes. The three options are summarized in Table 11.7, while further details for options (ii) and (iii), comparable with those in Table 11.6, are listed in Tables 11.8 and 11.9.

Table 11.7. A comparison of rough set analysis of the three options.

Option	Reducts	Core	Number of exact rules
(i) 43 objects	{1,8,12,13}, {1,8,11,13}, {1,6,8,13}, {1,3,7,8,13}, {1,2,8,13}	{1,8,13}	18
(ii) 28 objects	{1,8,12},{1,6,12}, {1,8,11},{1,6,8}, {1,3,8},{1,2,8}	{1}	11
(iii) 13 objects	{1,12},{1,11}, {1,6},{1,3},{1,2}	{1}	7

Table 11.7 also reveals that, with a smaller sample the reducts become smaller - in other words, theories require fewer factors to explain variation in the dependent variable. The core is reduced to a single factor in options (ii) and (iii), so that the sample size is the only essential factor in explaining the variation in the value of life estimates. In addition, in option (iii), from the set of reducts it can be inferred that variables 2,3,6,11 and 12 are explanatory factors, whereas the core variables 8 and 13 have disappeared. This means that the number of variables in the case studies'

estimation, the number of risk variables, the number of interaction terms, and the database and risk variable upon which the studies were based are important. That variable 8 has disappeared entirely from the reducts in option (iii) is explained by the fact that only those studies were considered here which had a t-statistic larger than 1.8 of the risk effect estimate. Similarly, variable 13 disappeared from the reducts in options (ii) and (iii) as it was a criterion for selection of cases in these options.

Table 11.8. Accuracy and quality of the classification of the decisional variable under option (ii) (number of objects = 28).

Class of decisional/dependent variable	Accuracy	Lower approximation	Upper approximation
1	1	10	10
2	0.7143	10	14
3	0.5	4	8
Accuracy of classification	0.75		
Quality of classification	0.8571		

Table 11.9. Accuracy and quality of the classification of the decisional variable under option (iii) (number of objects = 13).

Class of decisional/dependent variable	Accuracy	Lower approximation	Upper approximation
1	1	1	1
2	0.556	5	9
3	0.5	4	8
Accuracy of classification	0.5556		
Quality of classification	0.7143		

The results seen in Tables 11.8 and 11.9 include only 3 classes (as compared to 4 in Table 11.6), which is the result of (coincidentally) leaving out all cases or observations which had a value of the dependent variable that belonged to class 4 (defined as values above US$ 8 million). The accuracy and quality of the classification improves from option (i) to (ii), but deteriorates from either (i) or (ii) to (iii).

Other information can also be extracted from the rough set analysis, relating to the specific rules, and to the support of rules by cases or objects. If a rule is supported by more objects, than it is more important or relevant, for instance, in summarizing or aggregating the different single case study results, or for predictive purposes.

Finally, a sensitivity analysis can be performed, notably by changing thresholds or classes. Since variable 1, sample size, belongs to the core in all three options, a sensitivity analysis was performed both with regard to the classification of the dependent variable, i.e. value of life estimate, and the sample size. The sensitivity test, however, is only done for option (iii) as this contains the most consistent data base.

In the context of sensitivity analysis with respect to classification or coding of the explanatory variables, the focus is on the sample size, because it is the only variable that is an element of the core under all three options. With option (iii), involving 13 cases, extending the sample size classification by splitting up the class [5000-10000] into two classes [5000-5700] and [5700-10000] leads to a core and only one reduct equal to {1}. The reason is that variable 1 is represented by too many classes.

Reducing, on the other hand, the number of classes for the same variable from 6 to 4 (or, alternatively, the number of thresholds from 5 to 3), resulting in classes [0-1000], [1000-5000], [5000-10000] and [10000-∞] leads to a core of {1} and 11 reducts with 2, 3 and 4 variables: {1,11}, {1,12}, {1,2,10}, {1,3,10}, {1,6,10}, {1,2,4,7}, {1,3,4,7}, {1,2,5,7}, {1,3,5,7}, {1,4,6,7} and {1,5,6,7}. This larger number of reducts is because variable 1 is less able to explain the variation since it contains less classes. The accuracy and quality of the classification do not change from the original rough set outcome as shown in Table 11.9.

Sensitivity analysis with respect to the classification of the decision or dependent variable, also for option (iii), was done by adding two classes relative to the original classification. The original classification for the value of life was (in millions of US$): [0-1], [1-4], [4-8] and [8-∞]. The revision leads to classes [0-1], [1-2.5], [2.5-4], [4-5.7], [5.7-6.2] and [6.2-∞]. The core is then {1,2,5,6} while there are 4 reducts {1,2,5,6,7}, {1,2,5,6,10}, {1,2,5,6,11} and {1,2,5,6,12}. The accuracy and quality of the classification have not changed. This can be explained as follows. Adding classes for the independent variable implies that the coded information for this variable is more varied. Explanation of the variation in this variable, therefore, requires at least as many variables as with the original classification.

11.5 Comparison and Conclusions

The application of both statistical and rough set analysis to the three options yields results which to some extent contain differences. This is partly explained by the large number of regression models that can be estimated. It is difficult to make a choice in this case, and for this reason the significance of variables are given for some regressions. In any case, the sample size, the t-statistic of the risk effect and the subset of the database are essential in many of the outcomes, for both the statistical and the rough set analyses.

The rough set analysis leads to information about groups of factors that can uniquely determine the variation in values of life estimates and indicates which are essential variables in this respect. Such a clear-cut distinction is not possible with regression, where only significant variables can be identified.

The results of applying meta-analysis techniques to the hedonic wage studies are as follows. Three samples were analysed. The first is based on all model estimations within and between different studies, including 43 cases. The average value of life estimate turned out to be US$ 3.59 million in this case. Eight OLS regression were performed, with adjusted R^2 ranging from 0.28 to 0.59. The sample size, R^2, the number of variables, the t-statistic, the database and the linearity of the model were significant. In some estimations, however, only the t-statistic of the risk effect turned out to be relevant. In the other two samples studied, comparables databases and significant risk-coefficients are the criteria. Focusing only on the first of these, a sample of 28 cases results, with an average value of life estimate equal to US$ 2.34 million i.e. approximately one third lower than for the whole sample. In this smaller sample variables are much less correlated than in the whole sample. The most important change in the set of significant variables from the larger to the smaller sample is the addition of the indicator for risk. Using the second criterion, to restrict the sample even more by including only significant estimated risk coefficients, a sample of 13 cases remains, increasing correlation among variables somewhat relative to the second sample of 28 cases. The average value of life estimate is now more than two times higher than in the second sample, namely US$ 3.86. Additional significant variables are the number of risk variables and the Number of interaction terms in the models. Taking the position that the latter sample is most reliable, then a relatively high value of life results.

Similarly, the rough set theory was applied to all three samples. The smaller the sample, the smaller also the number of variables in the reducts, which may be regarded as specific sets of factors offering a full explanation of the phenomenon. The core variables are in the large sample the sample size, the t-statistic of the risk effect, and the dummy for the subset. The latter two imply that consideration of the smaller samples is useful, as these are constructed on values of these two variables. In these smaller samples the core consists only of the sample size, while in the smallest sample the other relevant variables are the number of variables, the number of risk variables, the number of interaction terms, the indicator for the database, and the indicator for risk. Sensitivity analysis does not change these outcomes. In conclusion, the results of statistical and rough set analysis tend to point in the same direction.

CHAPTER 12 EFFECTIVENESS OF PESTICIDE PRICE POLICIES IN AGRICULTURE

12.1 Introduction

The intensive use of pesticides in agricultural production in large parts of the world has been one of the key factors contributing to increased yields. There is an increasing concern, however, about the impact of pesticides on human and animal health and the environment more generally. As a result, several national and international initiatives have been taken to reduce the use of pesticides. The Commission of the European Communities (1992) has, for example, declared in its Fifth Environmental Action Programme that a significant reduction of pesticide use per unit of land should be achieved by the year 2000.

The major challenge for an EU pesticide policy is, in addition to the formulation of legislative compulsory measures, to find ways of stimulating farmers to adapt the less ecologically damaging farming practices. Various policy options exist for such a strategy such as, financial compensation for additional protective measures by farmers, the creation of a new market incentive by environmental quality labels and the direct regulation by licenses and certificates of competence for pesticide retailers and farmers (see Petterson *et al.*, 1989; Reus *et al.*, 1994; Verhoeven *et al.*, 1994; Wossink and Renkema, 1994).

In this chapter the policy options open to regulate pesticide prices are examined. This is done by looking into the behavioural response of farmers to levies on the use of pesticides. Two steps are taken. First, data on price elasticities are collected from the existing literature. Second, meta-analysis is deployed in order to investigate the determinants of pesticide price elasticity.

12.2 The Studies Examined

The aim of our literature search was the collection of empirical evidence on existing price elasticities of the demand for pesticides by farmers. Initial examination of the literature revealed that while there is considerable published work on such things as laboratory tests measuring the impacts of pesticides there are few studies of policy implications. In particular the literature on the behaviour of farmers regarding the use of pesticides is scant.

The research on the impacts of changing pesticide prices on pesticide use mainly consists of studies using damage-threshold models, econometric models and linear programming models. In most of these studies pesticides are included as one of the production factors. Sensitivity analyses regarding the specific impacts of changing pesticide price on pesticide use are therefore partial analyses of broader models.

Damage-threshold models indicate the level of crop disease at which it becomes cost-effective to apply a treatment by pesticides. To be able to work out the damage threshold, the product prices and the cost of pesticides must be known. These models can be used to assist farm management.

Econometric approaches use models composed of quantified relationships between the key variables of interest such as the price of pesticides, demand for pesticides, demand for other variable inputs like energy and seed, supply of output and demand for capital. Production behaviour of farm households is often modelled starting from neoclassical theory. All relationships, therefore, within these models have a causal interpretation. The models are in most cases quantified by applying statistical techniques to historical data (Oskam *et al.*, 1992). Their advantage is that they represent real behaviour. Their disadvantage is that new techniques cannot be incorporated.

Linear programming models are frequently employed to simulate the economic decision-making process at the micro level or farm level. Environmental aspects of agricultural production can in most cases also be incorporated for two reasons. First, linear programming provides an explicit and efficient optimum-seeking procedure. Second, once the program has been formulated, the results obtained by changing variables can rapidly be calculated. For the investigation of pesticide price regulation effects, the individual farm is a good starting point, because it is at this level that the actual decisions are made about cropping patterns, production intensities, use of input factors and so on.

The major advantage of using linear programming models for environmental economic research is that several activities producing the same product can be considered at the same time. The environmental impacts are represented in the technical coefficients of the process in the linear programming matrix. The objective functions include financial results reflected in gross margin figures. With this range of cropping variants, it is possible to investigate the effects of levy systems on pesticide use.

In this context, a major source of references is a study by Oskam *et al.* (1992), in which several recent economic and econometric analyses of possible pesticide price policies in four EU countries, along with some other separate studies, were reviewed. In total 13 studies were found which yielded empirical evidence on the price elasticity of the demand for pesticides. This may mean a bias towards North-European studies, but since pesticides are a major concern in the intensive North-European agriculture, the comparative meta-analytic investigation is still warranted as it may pinpoint key factors which might be further tested in a broader set of more representative studies (by consulting e.g. World Agricultural Economics or Rural Sociology Abstracts). Some studies which provided information on the relation between levies on pesticides and the use of pesticides, did unfortunately not express these relationships in terms of price elasticities (the studies of Jong (1991) and Wossink (1993)). The following studies were of use.

1. Sweden

Damage-threshold models were used by Kungl. skogs- och lantbruksakademien (1989) to establish damage thresholds for various crops, fungus diseases and insect infestations. Calculations were done on data gathered from a large number of experimental fields in Sweden. Using damage-threshold models, pest control for a certain crop/disease combination was investigated for economic feasibility.

Quantitative scenarios were constructed with different quotas on pesticides. It was assumed that producers would receive compensatory payments for income losses incurred from the imposition of pesticide quotas. The compensation could take place by adjusting product prices. The result would be an increase in domestic prices for farm products.

An interregional, spatial linear programming model of Swedish food production was then used to elucidate the aggregated economic effects of reduced pesticide use. From this the sensitivity of pesticide use to price change was derived by assuming that imposing a levy equivalent to the value of the additional revenue obtained by additional treatment would bring about the targeted reductions in pesticide use.

2. Denmark

This study of Dubgaard (1991) also used an ad-hoc damage-threshold model with the results interpreted on an aggregated level. Data were collected from experimental fields throughout Denmark. To determine induced technological change in the model, estimates made by experts in crop protection were used.

3. Denmark

Study (2) also used econometric methods to investigate the observed behaviour of farmers to obtain reference material for Dubgaard (1991). A regression analysis was carried out based on annual Danish data on prices and pesticide use for the period 1971-1985. In the regression equations the average number of treatments per hectare were reflected as a function of index prices of pesticides and a trend variable.

4. Germany

Schulte's study (1983) looks at the effects of reduced application of fertilizers and pesticides on farmers' incomes. Among the instruments examined was a tax on fungicides. Linear programming models were used for five farms, differentiating between factors such as farm size and region. The farms were three arable farms, an intensive livestock farm and a mixed farm.

5. Netherlands

In the study by Elhorst (1990) an econometric approach was used, namely an agricultural household production model, in which the neoclassical theory of producer and consumer behaviour was incorporated. Price sensitivities to fertilizers, pesticides and other variable inputs were investigated in Dutch arable farming. These inputs were combined under the heading 'non-factor inputs'. Data were used for the period 1980-1987.

6. Netherlands

Oskam's study (1992) contains research on the use of fertilizers and pesticides in Dutch arable farming, using data from a stratified sample of arable farms for the period 1970-1987. A distinction was made between specialized farms, and farms where 50-80% is reserved for arable farming. The study reveals some factors explaining the use of pesticides, including the cropping plan of the farms and the quantity and quality of crops.

Price sensitivities of different crops to the price of pesticides were determined, in first instance, by an analysis that first estimated the volume of pesticide use per crop per year. Conducting regression analysis, these estimates were then used to determine the extent to which pesticide use is increasing. In addition, pesticide use is influenced by the prices of crops and pesticides. The resulting price elasticities are medium-term elasticities because the specified relationship between pesticide use and pesticide price incorporates short-term as well as long-term effects.

7. Netherlands

The aim of the study of Houwen (1993) was to assess the effects of the developments in the EU agricultural policy and the strengthening of the environmental policy on arable farms in part of the Netherlands. It used a linear programming model at a micro level, based on environmental-economic models.

8. United States

In the study of Chambers and Lichtenberg (1994) an econometric model was used to estimate pest damage. The basis was a framework for estimating pesticide technologies that recognizes the peculiar role pesticides play as damage-control agents. A multi-output version of such a model was developed and the generalized model's dual representation derived. The econometric procedure for estimating the dual technology was applied to an aggregate time-series data set for the entire U.S. agricultural production sector for 1949-1990. Three separate econometric estimate procedures were used: an exponential model, a logistic model and a generalized-Leontief model, providing three different estimates of pesticide price elasticity.

9. Netherlands

Jong (1991) used an environmental-economic linear programming model on a micro level to investigate, from a business-economic perspective, the size of levies on pesticides to achieve the reduction goals of the Dutch multi-year crop protection plan. The impacts of pesticide levies on the environment and the income of farmers were, therefore, investigated. Representative models of large farms were formulated for four specific agricultural regions, while two representative models of small firms were selected, resulting in a total of six different models of agricultural farms. To obtain data on a national level, the totals of the six types of linear programming models were aggregated.

The study did not have explicit values for pesticide-use elasticities. A quantitive assessment of farmers' change in use is given in Table 12.1. Initial prices are, however, not known. This lack of information makes the results of limited use for our analysis.

10. Netherlands

The study of Oskam *et al.* (1992) elucidated the influence of pesticide prices on the demand for pesticides in Dutch arable farming and in Dutch horticulture using econometric models. The models assumed short-term profit-maximizing behaviour with respect to variable input factors. Both models were aggregate models, covering respectively the Dutch arable and horticulture sector as a whole. They were quantified using data for the period 1970-1988. Short-term elasticities (one year) were assessed as well as medium-term (three years) and long-term (ten years) elasticities. It was assumed that in the medium term the amounts of capital goods will also partially adjust to a change in the pesticide price, influencing in turn pesticide demand. In the long run it was assumed that the amounts of capital goods will fully adapt to changed pesticide prices.

Table 12.1. Results of the LP model used by Jong (1991).

Pesticide	Levy in Hfl per kg a.i. and % change in use (related to current situation)							
	0	10	25	50	75	100	150	200
Herbicides	- 22	- 58	- 62	- 71	- 88	- 89	- 91	- 91
Insecticides	- 27	- 37	- 39	- 43	- 50	- 52	- 56	- 56
Fungicides	+ 6	- 24	- 27	- 31	- 39	- 55	- 63	- 63

Note: Figures have been derived from Appendix VII "Results in tables and graphics".
Source: Jong (1991).

11. Netherlands

Aaltink (1992) used an econometric approach to investigate pesticides' price elasticities. He based his model on a profit function as well as a cost function with the latter assuming that the amount of production is fixed in the short term. The same data set was used for quantifying both models, but prices and profits were normalized by the price of the other variable input in the case of the profit-function model.

12. Netherlands

The study of Wossink (1993) used a micro level linear programming model for the evaluation of various scenario developments regarding changing price and policy conditions for the arable farming sector. The model enabled the changes in farm organization resulting from six scenarios to be ascertained. These scenarios represented combinations of technical developments, environmental regulations and different forms of price and market policies. The relevant scenario for the assessment of the impact of pesticide levies was calculated for the year 2000, implying long term elasticities.

The model was quantified by applying it to a series of eight representative farm types in a specific region in the Netherlands. These were selected using cluster analysis of specialized arable farms in this region.

Just as with Jong (1991), no explicit values were assessed for pesticide price elasticities. In Table 12.2 quantitative information is given of the farmers' change of use of pesticides under 5 different levy scenarios. To make these results usable for both kinds of pesticides, the geometric average of these relative changes is calculated in order to approximate the elasticity concerned.

Table 12.2. Results of the LP model used by Wossink (1993).

Pesticide	Fixed levy in Hfl per kg a.i.				
	10	25	50	55	60
Herbicides					
% price change	+ 22	+ 55	+ 110	+ 121	+ 132
% change in use	- 37	- 40	- 68	- 68	- 77
Insecticides					
% price change	+ 2	+ 5	+ 10	+ 11	+ 12
% change in use	+ 12	- 2	- 3	- 3	- 26
Fungicides					
% change in use	- 28	- 33	- 33	- 33	- 34

Note: Figures have been derived from Tables VII.1 and VII.5. Initial (average) pesticide prices were calculated on the data in Table VII.1 of Wossink (1993). These were respectively for herbicides Hfl 45.24 and for insecticides Hfl 488.56. For fungicides it appeared impossible to estimate relative price changes.
Source: Wossink (1993).

12. Philippines

Antle and Pingali (1994) used an econometric approach to investigate the impacts of pesticide use on farmer health by means of a health model and, subsequently, the impacts of farmer health on farm productivity by means of a production model. These models were then joined in a simulation analysis to investigate the health and productivity tradeoffs implied by a policy to restrict the use of pesticides, including the option of a price regulation for insecticides and herbicides.

A distinction was made between the direct regulating effect of pesticide price policy on farm productivity and an indirect effect, through the impact on health. This indirect effect thus differs from the other econometric studies, that do not account for health effects. These health effects are, however, expected to be less important in developed countries because of more advanced spraying materials and stronger laws regarding the handling of pesticides.

The study concentrated on the rice producing sector in two regions. The empirical base for the estimation of the production model contained data from a farm-level survey in both regions, while the estimation of the health model included data on pesticide handling during the period 1987-1990 and 1988-1991.

12.3 The Application of Meta-analysis to Pesticide Price Policy Studies

The estimations of elasticity values in the studies range from practically zero (-0.05) to a level of -1.53. The number of studies, however, is not sufficient to use any conventional statistical estimation method to investigate the importance of different factors. Rough set analysis offers an alternative. The following set of hypotheses are used for the selection of potential explanatory factors.

1. Country. It can be hypothesized that elasticities vary between countries. Various reasons may explain this. First, the type and quality of agricultural soils differ between various countries. Similar crops in different countries may have different growing characteristics. This would mean that the marginal impacts of using pesticides may differ, resulting in different levels of use. Second, the agricultural policies may differ between countries, e.g. in the area of legal quality requirements to agricultural crops. Such requirements are not only dependent on the country's own policy but also on policies of countries to which the crop is being exported. Such policies place different levels of restrictions on agricultural production, resulting in the demand for pesticides not having a monotone relationship with pesticide costs, but rather a stepwise relationship, specific to each country. Third, the influence of farmers' income should be taken into consideration. It can be expected that pesticide use will be lower as the average income of farmers is lower. It would be important here to make a distinction between developed countries and third world countries in this sense.

2. Data collection period. A second important factor may be the time period over which the data for the studies were collected. The length of these periods can be regarded as a proxy for the quantity of the data set used. Second, the absolute point in time is also relevant, since price elasticities may vary over time. Between the first and last year of collection in our data set, there may be large fluctuations caused by socio-economic and technical developments.

3. Year of publication. The year of publication may be a good indicator for the period in which the research was carried out. This aspect may be important, as developments in data analysis techniques and improvements of the quality of economic models over time may reduce the level of errors in the research results (see Van den Bergh *et al.*, 1995b).

4. Publication channel. The quality of research may be related to the channel of publication. In this sense, it is useful to distinguish between publications media, e.g. M.A. thesis, dissertation, monograph, contribution to a book or journal article. Furthermore, there may be a difference between independent research and commercial research. Clearly, in principle, such differences should of course not influence the results, but practice may be different (see Baaijens, 1996).

5. Type of agriculture. Pesticide price elasticities differ considerably between types of agriculture. Each type of agriculture has its specific crop species, by which available alternatives for crop protection are specific to each type of agriculture. It is to be expected that the fewer alternatives there are, the smaller the price elasticity will be. Furthermore, each type of cropping activity has its specific share of pesticide costs in total costs. If this ratio is high, a rise in pesticide costs will have a relatively large impact on total costs, probably resulting in a relatively high price elasticity.

6. Pesticide type. It is to be expected that cost elasticities differ among different types of pesticides. This is due to the different natures of plagues occurring in agricultural activities. The demand for pesticides against plagues which cause total destruction of a harvest is probably totally price-inelastic. Specific plagues may, however, ravage a variety of crops in different manners. It is difficult, therefore, to assign to each type of pesticide a certain level indicating its potential to prevent or reduce crop damage.

7. Levy. In conformity with micro-economic theory, the price elasticity is expected to vary according to the level of the respective price. This means that one would not expect to find a uniform elasticity across countries which have different pesticide price levels.

8. Model. A factor differentiating studies is the type of model used to estimate cost elasticities. Here a distinction is made between linear programming techniques and econometric methods.

For the application of rough-set analysis the list of variables is coded as in Table 12.5. Four alternative codifications are presented and used in four respective 'rounds' of analysis. In the first round, the following classifications were made:
Country: a distinction is made between each single country.
Data period: there are 4 classifications of time periods lasting until approximately 1990; their length roughly doubles with each class (5, 10, 20 and 40 years respectively). This means that the classification is consistent between period length and the absolute point in time.
Year of publication: research carried out until 1990 is distinguished from research carried out after 1990.
Publication channel: the following classification was used: research reports, journals (A means a top-class journal, E a lower classification), dissertations and theses.
Type of agriculture: a distinction is made between farms with a 100% arable activity and farms with mixed activities. Such a classification is a very rough one, and it does not discriminate between specific crop species. The data base, however, does not allow for a better classification, because most of the studies concern farms growing a broad range of crop species.
Pesticide type: 7 classes of combinations of pesticide types are used: herbicides, fungicides and insecticides as single applications, the combinations of two different pesticides (herbicides/insecticides and fungicides/insecticides), and the mixed use of all kinds of pesticides with or without fertilizers.
Levy: Studies that estimate cost elasticities based on a specific initial price level of pesticides are distinguished from those that do not.
Model: two modelling types are taken into consideration: linear programming models and econometric models.

The running of the data in a rough-set analysis resulted in the set of 9 'reducts', shown in Table 12.5. This relatively large number of reducts is not very satisfactory. The core set of attributes, however, consisted of the single variable reflecting the type of agriculture, which is revealing in terms of the relative explanatory power of this one variable in relation to the others.

Table 12.5. Coding of attributes in four rounds.

	R1	R2	R3	R4		R1	R2	R3	R4
A1. Country					*A6. Pesticide type*				
Sweden	1	1	1	1	All	1	1	1	1
Denmark	2	1	1	2	All incl. fertilizers	1	1	1	1
Germany	3	1	1	3	Fungicides	2	2	2	2
Netherlands	4	1	1	4	Herbi/insecticides	3	3	3	3
USA	5	1	1	5	Fungi/insecticides	3	3	3	3
Philippines	6	2	2	6	Herbicides	4	4	4	4
					Insecticides	5	5	5	5
A2. Data period									
1971-1989	1	1	1	1	*A7. Levy*				
1970-1987	1	1	1	1	Certain levy	1	1	-	-
1970-1988	1	1	1	1	No certain levy	2	2		
1980-1987	2	2	2	2					
1949-1990	4	4	4	4	*A8. Model*				
1987-1991	3	3	3	3	Linear programming	1	1	1	1
Unknown	5	5	5	5	Econometric	2	2	2	2
A3. Year of publication									
Until 1990	1	1	-	-					
After 1990	2	2							
					Price Elasticity				
A4. Publication					0 to -1.9	1	1	1	1
channel					-1.9 to -0.3	2	2	2	2
Research report	1	1	-	-	-0.3 to -0.6	3	3	3	3
Journal	2	2			-0.6 to -1	4	4	4	4
Dissertation	3	3			-1 to -infinity	5	5	5	5
Thesis	4	4							
A5. Type of agriculture									
Arable									
Mixed	1	1	1	1					
	2	2	2	2					

Notes: Ri denotes round i, i=1,2,3,4. "-" means that the respective variable is not included.

A second round employed the classification of the country variable to only two classes: for developed countries and developing countries. This distinction is meaningful with regard to the influence of the farmers' income on the use of pesticides. Rough set analysis run on the adjusted data matrix produces then only 5 reducts, while the core set consists of two variables: type of agriculture and of type of pesticides. Across the studies, therefore, there are two key variables that excercise a relatively strong impact on the use of pesticides.

The third round represents another option, namely to omit the following variables: year of publication, publication channel and the levy variable. The classifications of the other variables are retained. The results of this round lead to only one reduct of four variables (Table 12.5) which are, in this case, equal to the core set. In addition to the type of agriculture and the type of pesticide, the core set consists of the data period and the model type variables. Apparently, the two attributes of the set at hand that are related to the basic methodological and empirical differences across the studies, have, in conjunction with variables defining the type of agricultural operation, the strongest explanatory power for the variance in research outcomes. With regard to the data period variable, it is not, however, clear which of its dimensions, the length of the data collection period or the date of this period, cause this variation.

Table 12.6. Reducts and cores.

	R1	R2	R3	R4
Reducts (or minimal sets)	1. {4,5,6,8} 2. {2,5,6,8} 3. {1,4,5,8} 4. {1,2,5,7,8} 5. {3,4,5,6,7} 6. {2,5,6,7} 7. {1,3,4,5,7} 8. {1,2,4,5} 9. {2,4,5,6}	1. {4,5,6,8} 2. {2,5,6,8} 3. {3,4,5,6,7} 4. {2,5,6,7} 5. {2,4,5,6}	1. {2,5,6,8}	1. {2,5,6,8}
Core set	A5	A5, A6	A2, A5, A6, A8	A2, A5, A6, A8
Accuracy of classification	0.7778	0.7778	0.7778	0.7778
Quality of classification	0.8750	0.8750	0.8750	0.8750

Finally, the robustness of the above results is tested. This is done by taking into consideration the same variables as in the third round of analysis. The coding of the country variable is changed back to its original format, to investigate whether the variable would consistently fall out of the core set. This analysis leads to exactly the same reduct as in the third round. The conclusion, therefore, is that in comparison with other possible explanatory factors, price elasticities of pesticide demand do not show large variations that can be explained by geographical factors.

In addition to the investigation of the explanatory power of the various variables, we can also look at the direction in which they affect research outcomes on the price elasticities of pesticide demand. Decision rules are, therefore, examined. These can be expressed as logical 'if ...., then ...' statements that can be derived from the used information table. They can be used for predictions by matching the descriptions of cases to be predicted with the decision rules.

The rules, based on the attributes of the core set of the four attributes which appear to be powerful turn out to be the same in each of the four classification rounds. These rules are shown in Table 12.7.

Table 12.7. Set of decision rules, based on the core set of attributes A2, A5, A6 and A8, generated for all rounds of analysis.

Number	Rule (attributes)	Class (price elasticities)
1	A2=5 and A5=1 and A6=1 and A8=2	1 or 3 (non-deterministic rule)
2	A2=2 and A6=1	2
3	A2=3 and A6=3	2
4	A2=5 and A5=2 and A6=1	2
5	A2=1 and A5=2 and A6=1	2
6	A2=5 and A5=1 and A6=1 and A8=1	2
7	A6=2	3
8	A2=1 and A5=1 and A6=1	3
9	A6=4	4
10	A6=5	4
11	A2=1 and A6=3	4
12	A2=4	5

We can draw some quite interesting conclusions from this result. First, it appears that when the period of data collection is longer, the study leads to a relatively higher price elasticity. A data period of 10 years leads to a significantly lower elasticity than a data collection period of 40 years. This is a plausible result, since in the long run, price elasticity of pesticide demand is greater than in the short run because of the long-term adaptive behaviour and flexibility of farmers (the Le Chatelier principle; see Varian, 1992). It should be noted however, that this statement would need some further research, as there might be a difference between a long period of observation and the time span used in the model specification. Second, it seems that when a farm uses the full range of pesticides, it is less price-elastic than when it uses a lower number of pesticide types. This risk dispersion result can perhaps be explained by a partial substitutability between different pesticide options, but also this result calls for some caution, as it is not always clear from the studies that farmers are only using one type of pesticide, when the research focuses on that pesticide (De Wit, 1992). In the cases of the other two variables (type of agriculture and type of research model), no clear inferences can be made regarding their influences on the level of price elasticities.

12.4 Conclusions

Existing evidence on the size of farmers' price elasticities of pesticide demand is fragmented. Results of the various studies which have been computed fluctuate between an elasticity of almost zero and -1.5. At the policy level, it is, however, important to have some insight into the factors which cause the differences in price sensitivity in order to develop optimal pest control policies. By using a form of meta-analysis, one can detect the most important factors that influence variations in research outcomes. In this way, relevant policy information can be generated on differences in sensitivity between various farm segments, in terms of type of agriculture, type of pesticides used and so on, as well as information helpful for isolating future research topics. Some intuitive expectations were more or less confirmed by the findings of this analysis. It appears that in conjunction with variables defining the type of agriculture and type of pesticides used, variables that define the basic methodological and empirical differences across the different studies have strong explanatory powers in accounting for variances in research outcomes.

Furthermore, the results confirmed the expectation that in the long run the price elasticity of pesticide demand is higher than in the short run because of the adaptive behaviour of farmers. Second, it seems that when a farm uses the full range of pesticides, it is less price-elastic than when it uses a smaller number of pesticide types. This suggests that the level of substitutability or 'packaging' between different pesticide options is a critical factor.

CHAPTER 13 EFFECTIVENESS OF TRAFFIC RESTRAINT POLICIES

13.1 Introduction

Public policy regarding the urban transport problem has undergone important changes over the past thirty years. From one perspective the perception of the urban transport problem has widened, as issues such as environmental protection, safety and and problems in financing public transport provision have been added to the more traditional concern of excessive traffic congestion. Nevertheless, traffic congestion remains at the forefront of policy analysis, although here too there has been a change in perspective, in this case in relation to the appropriate emphasis public policy should take. In the past, policy has often centred on expanding the infrastructure for traffic, adopting technologies which increase the capacity of the existing network and on attempting to make public modes more attractive. The shift has recently been towards containing traffic growth by discouraging the use of the private car either through physical restraints or by the adoption of fiscal measures.

The types of traffic restraint measures used vary and often differ in their detailed objectives although containment of traffic congestion is still generally at the forefront of thinking. Equally there is a growing body of knowledge as to the effectiveness of alternative strategies of traffic restraint as more academic, consultancy and official analyses are completed.

Nevertheless, and despite the emergence of a number of survey papers, there has been little effort to systematically examine the effectiveness of the different policies in a rigorous, statistical manner. The task of doing so is not an easy one and this case study in no way pretends to produce anything other than a preliminary set of results. It does seem rather wasteful, however, to expend considerable resources on conducting fresh studies of potential traffic restraint policies for each city or circumstance when there is already quite a significant body of information lying around which can, at the very least, be sifted for important initial insights.

The aim of this chapter is to examine potential insights which meta-analysis can offer in the field of quantitative transport economics. The simple meta-regression approach employed here brings together the results of a number of studies which have explored the implications of various urban traffic restraint schemes on congestion levels. From the longer term perspective, more importantly the chapter seeks to use the results and experiences of this meta-analysis work to suggest ways in which the results of empirical studies may usefully be presented to allow more detailed comparative analysis to be completed.

The chapter continues by employing meta-regression analysis to explore the effectiveness of different traffic restraint instruments. It is perhaps appropriate to pre-empt one of the conclusions of this work to say that the results offered here are limited by the lack of consistency with which the results of transport studies are reported. Greater standardisation in the way output is presented, as is common in other areas of study such as medicine and the physical sciences, makes an overall assessment of policy impacts much easier. The final section looks at the longer term potential for employing meta-analysis as an instrument of urban transport policy appraisal.

13.2 A Simple Meta-regression Analysis

To illustrate the potential for meta-analysis and, at the same time, highlight some of the limitations of conducting this type of work, given the nature and output of existing case studies, the results from a number of traffic restraint studies, covering a variety of different policies and countries, are subjected to a meta-regression analysis at a meso-level.

In an attempt to encapsulate the range of possible influences on the effectiveness of policies on reducing congestion (RC), allowance is made in our meta-regression for five broad categories of effect which may have influenced the outcome of the case studies examined:

- the macro-geographical area in which the work was completed (C) - essentially this was included to pick up differing socio-cultural attitudes to transport;
- whether the analysis was at the regional or urban level (L) to provide an indication of the level of spatial coverage of each case study;
- the nature of the traffic restraint policy instrument deployed (M);
- the level of any public transport improvement (I) to reflect complementary policies; and
- the type of vehicle affected by the policy (T).

The overall formulation of the meta-regression model explored, therefore, is one taking the general form:

$$RC = f(C, L, M, I, T) \tag{13.1}$$

This general framework is applied to the results of a series of studies that have attempted to assess the implications on traffic congestion of a number of policy options. The case studies examined were extracted from: Cameron (1991); Cheslow (1978); Cracknell *et al.* (1975); Elliot (1975); Elmberg (1973); Fowkes *et al.* (1993); Ghali and Smith (1994); Gomez-Ibanez and Fauth (1980); Hartmut and Pharoah (1994); Hau (1990); Larsen (1988); May (1975; 1992); May and Nash (1990); Miller and Everett (1982); Ministerie van Verkeer en Waterstaat (1988); Rathbone (1992); MVA Consultancy (1995); Spielburg (1978); and Storstadstrafik (1989). Since the aim of this study is as much to explore the usefulness of meta-analysis as it is to produce original findings, the studies were collected simply by making use of a university catalogue and, hence, may not entirely be random either because of subjectivity in the initial collection by the library or because of idiosyncrasies on the part of the cataloguers. Many case studies were also omitted, because they did not contain sufficient detail to make their inclusion possible. This chapter employs 22 case study examples of the ways in which policies may reduce traffic congestion. The meta-regression was run in MICROFIT using ordinary-least squares. Given the constrained nature of the dependent variable (percentage reduction in traffic), strictly a non-linear transformation should have been done, but since the results are not intended for forecasting purposes this was omitted.

The details of the parameters and their associated t-values obtained are set out in Table 13.1. The meta-regression embraces the following independent variables:

HIG: Dummy variable taking unity value for Holland, Italy and Germany.
SCAND: Dummy variable taking unity value for Scandinavian countries.
SINGHK: Dummy variable taking unity value for Singapore and Hong Kong.
USA: Dummy variable taking unity value for USA.
REG: Dummy variable taking unity value for regional based analysis.
SLS: Dummy variable taking unity value for supplementary licensing policy.
TZ: Dummy variable taking unity value for traffic zoning.
CP: Dummy variable taking unity value for congestion pricing.
CRC: Dummy variable taking unity value for car restricted city.
IPS: Dummy variable taking unity value for improvements in public transport
CAR: Dummy variable taking unity value for policies applying only to cars.

Table 13.1. Estimated coefficients derived from the meta-regression.

Variable	Coefficient	t-value
HIG	-28.243	-2.426
SCAND	17.509	2.959
SINGHK	23.293	2.879
USA	11.503	3.008
REG	8.667	1.286
SLS	5.991	1.346
TZ	13.725	1.765
CP	14.469	3.720
CRC	75.807	5.582
IPS	15.106	3.446
CAR	-15.171	-4.338

The results are far from statistically ideal but they do provide some useful general insights into the effects of traffic restraint policies. The estimated model contains no constant; the adjusted R-squared generated is 0.786. There seems, for example, to be fairly strong evidence from the significance of the country/macro-area related dummy variables that there are important national differences in the way traffic restrain measures influence congestion. Equally, there are differences between the impact of the various types of restrain measures themselves. The evidence of studies which look at congestion pricing indicates that it would have a significant effect on the degree of traffic reduction.

In terms of individual policy measures, those applied purely to cars seem to be counterproductive, although contrary to this, and possibly because such policies are anticipated to be more coordinated, car restricted cities would seem to provide a useful method of reducing traffic congestion. Equally, road pricing policies seem to offer a powerful traffic restraint policy, although recent debates in many countries have highlighted the problem of introducing them. Also the analysis is not capable of comparing the implications of the various forms of road pricing which are possible.

The aim of this core study, however, is less to dwell on the output of this work, which is designed to offer a flavour of meta-regression methods, than to comment on its limitations and offer some suggestions as to ways in which meta-analysis may be developed to be of more practical use in transport analysis.

What is missing from these results, however, or more correctly what really limits their real usefulness, is the way the analysis has by necessity been conducted. In particular, there is a heavy reliance on dummy explanatory variables, and hence on ordinal judgements, in the meta-regression. This will always be unavoidable with some types of information which is either naturally discrete, such as the geographical origin of the case study, or is only available in qualitative form, such as success or failure. In other cases, however, because of differences in reporting methods it becomes necessary to revert to discrete variables or the use of estimation techniques which effectively treat them as ordinal inputs. Even dependent variables are often expressed in different ways; in the case of congestion reduction this may be quantified as higher speed, greater flow or enhanced volume of traffic.

In the medical sciences and physical sciences there tends to be rather more standardisation in the presentation of results, experimental methods used and other details which makes the conducting of meta-analysis more tractable. Added to this, in many cases important information, such as the statistical significance of parameters, is simply not reported. The lack of consistence in reporting the transport economics field is not out of line with that found in other areas of economics and is perhaps, in part, explained by the adversarial nature of the uses to which transport studies are put (e.g. into public inquiries). Consistence in presenting future findings would, however, not only facilitate the wider use of meta-analysis but also make more basic assessments less haphazard.

A second difficulty of using meta-analysis in the transport field at present has less to do with the demands of statistical techniques and more to do with information availability. Initially, there is the simple problem of obtaining details of a large, random sample of studies. Much transport economics work is also often published in non-economics journals and, in no small number of instances, in non-social science disciplines such as civil engineering which adds to the problem. While modern, computerised information retrieval systems now makes this slightly easier, many important consultancy or government reports often go unrecorded and the record of past studies is, in any case, inevitably far from complete. As time passes one would hope that these types of difficulty will diminish.

13.3 Conclusions

In this chapter one particular aspect of transport economics has been explored. The body of studies exploring the effectiveness of alternative traffic policies aimed at containing urban traffic congestion is immense and is growing. What meta-analysis offers is the prospect of statistically bringing together and drawing additional insights from the results of such case study analyses. The empirical work present in this chapter makes no pretence of being in any way definitive, indeed far from it. It is aimed at being illustrative of both the potential of meta-analysis in this field and, perhaps of more importance, the actual problems of applying to the types of case study work which have been completed to date. It seeks to illustrate what statistical synthesis analysis can offer and to provide some positive suggestions regarding the ways in which, by quite simple means, its use could fruitfully be extended in the future.

CHAPTER 14 IMPACTS OF MOBILITY AND TRANSPORT POLICY

14.1 Introduction

Transport does not only create congestion in many countries, but also imposes itself in many other ways on society. The environmental impacts range from local effects such as traffic noise, odour, vibration, lead, benzene, community severence and visual intrusion, through a range of transboundary effects such as nitrogen oxide (NO_X) emissions, to global pollutants such as carbon dioxide (CO_2). It is often seen as a major influence on location decisions and on the spatial distribution of economic development. Additionally, there are efficiency considerations within transport itself especially concerning sub-optimal levels of congestion which waste time and other resources.

The practical problem in taking these effects into account in economic decision making processes is that they are often outside of the price mechanism with the result that monetary values are not associated with them in the conventional way. In many cases the result is excessive utilisation of transport because the prices perceived by users ignore external costs. Equally, while attempts have been made to apply cost benefit analysis (CBA) to the appraisal of transport investment appraisal this is only of a partial nature with, for example, environmental costs treated in a quantitative manner or in index form rather than as monetary equivalents comparable with the financial costs of the infrastructure and vehicle operating costs.

This problem of how to handle such things as environmental costs and development implications of transport are not ones that are likely to go away in the near future. There are concerns, for instance, that Europe may be confronting serious development constraint because transport and communications networks have inadequate capacity and are excessively segmented. In terms of utilisation of the existing physical capacity, there has clearly been significant growth in transport movements over the immediate past. The Single Market initiative, by stimulating the growth of European Union members states and allowing freer movement between them, is likely to stimulate rapid growth in future intra-Union, international transport above the already high projections of domestic traffic growth made by individual European nations (e.g. UK Department of Transport, 1989b). This is at a time when pan-European policies aimed at creating a multi-modal European-oriented infrastructure network are only just beginning to emerge. See, for example, Commission of the European Communities (1988; 1992), for discussion of the EU situation. Button *et al.* (1994) offers a wider pan-European perspective.

The concern is to ensure that the existing networks, and the additions being made to them, are being utilised in a socially efficient manner. This raises, amongst other things, questions of appropriate pricing and taxation. Already some 40% of the designated European motorway network is tolled and there are measures in many countries, including privatisations, to align rail charges more closely to commercial costs. What is also occurring, albeit at a much slower pace, is an effort to make transport users more aware of the social costs, and especially environmental costs, of the activities. The adoption of the OECD polluter-pay principle, for example, puts the onus of policy making on making transport users cognisant of the wider implications of their use of transport infrastructure.

One of the problems which is becoming transparent is the lack of adequate methods for rigorously appraising major transport projects, let alone the contribution that they may make to wider network developments. Equally, for the polluter-pays principle to be implemented there is a need to know what the payment (be it an emissions charge or the cost of conforming to a standard) should be. Much of the on-going decision-making is based upon qualitative judgements rather than quantitative evaluation. Ideally, some form of network assessment technique is required to help handle the complexities of a holistic approach to investment and charging, but given both the reality of the extant piecemeal approach to transport analysis and the conceptual difficulties that exist in operationalising a suitable methodology (Hauser and Maggi (1993) illustrate some of the problems of devising a practical, network based framework) this is still some way in the future.

A major difficulty in the short term, and in the context of existing CBA tools, is the problem of placing environmental and development considerations on an equal footing with the more traditional user effects. This, for example, is a problem with the COBA procedure (UK Department of Transport, 1989a). While CBA was somewhat discredited in the 1980s, mainly due to problems of operationalisation, it seems to make a comeback in some countries, although some still have reservations about its validity (Adams, 1993). Accounts of valuation and appraisal techniques used in other industrialised countries are to be found in Barde and Pearce (1991).

Recent years have thus seen a significant upsurge of studies seeking to examine the wider impacts of transport. The outcome has been a wide range of results and conclusions. A number of extremely useful surveys have emerged that have brought together much of this work and offered critical assessment of the methods used and of the results generated. These surveys have, however, tended to be of a traditional literary nature and have often produced lists of findings and taxonomies of procedures employed.

The aim of this chapter is not to produce new valuations of environmental effects nor to offer original findings relating to other aspects of transport activities. Nor is it to offer a conventional survey of the literature in the field. Rather it explores the alternative analytical way of meta-analysis of bringing together the growing number of empirical studies concerned with the wider implications of transport in order to gain greater insights into the causes of the variations in results obtained (e.g. in relation to techniques used, country of study, context of work, etc.).

With respect to the methodological levels discussed in Chapter 3, the following issues are important for a meta-approach of environmental effects of transport. At the study level, we are dealing here with the above mentioned problem of a large degree of heterogeneity between the valuation methods used in impact studies. In addition, there is heterogeneity in measurement of environmental effects as a variety of environmental indicators are used across studies. A meta-approach must ideally deal with the variations in used valuation models and outcome dimension. Furthermore, at the study selection level, care is taken with regards to the inclusion of studies in the analysis as a wide body of impact studies exist which are eligible.

At the meta-analysis level itself, the technique of statistical analysis is chosen for each of the subjects presented in this chapter.

In Section 14.2, we will elaborate on problems of evaluating large scale transport projects. Then we proceed with the application of meta-analyses in various fields of evaluation of mobility impacts (Section 14.3). Section 14.4 contains general conclusions from these applications.

14.2 Problems of Evaluating Large Scale Projects

Forecasting the need for additional transport capacity and the ultimate consequences of building it have long been recognised as difficult. Foster (1968) pinpointed some of the major problems a number of years ago, *viz.*: "The forecasting of transport needs and the planning of transport investment should draw more upon the social sciences than most industrial planning and more even than most forecasting of public expenditure. The principal reason is that transport can rely less on the immediate practical aid to forecasting and planning given by market research." These difficulties persist despite the fact that a viable, if not ideal, basic CBA framework for analysis has been established for many years. This can at least be traced back to the Victoria Line (Foster and Beesley, 1963) and Third London Airport (Commission On the Third London Airport, 1972) studies but the underlying methodology predates these (see Mishan, 1988).

In terms of putting monetary values on the diverse implications of new transport infrastructure a variety of methods are now employed (see Figure 14.1). That it is legitimate to put money values on external costs of atmospheric pollution, noise, vibration and so on is not, of course, without critics. Strictly, economists do not try to value the environment but rather look at human preferences regarding differing states of the environment. To quote Pearce (1992): "There is, of course, the view that we that we cannot "value the environment". But the meaning of this objection is not always clear, and confusion has arisen because economists have themselves used slipshod language. What economic valuation does is to measure human preferences for or against changes in the state of environments. It does not "value the environment". Indeed, it is not clear exactly what "valuing the environment" would mean". This is not a debate, though, we go into but rather take the view that if CBA is to be used then it is sensible to gain as many insights as possible concerning the appropriate values to use and to as comprehensive as possible in coverage.

At the outset, it should be said that some of the problems encountered in using monetary values stem not from the accuracy of the monetary valuation techniques *per se*, but rather from shortcomings in the overall impact assessment process. Figure 14.1 shows there is a chain of events leading initially from failings in the economic system that may be due to market imperfections, but can also result from government intervention failures (Button, 1992), through distortions in transport supply and demand to excessive environmental degradation and, ultimately, the costs this imposes on society. A major difficulty is that even if an accepted monetary value could be derived for environmental impacts of say, an excess decibel of noise, there is the need to be able to relate that to traffic levels. A 5% error in valuation may have only a small effect on the overall accuracy if the traffic forecasts are extremely poor and the truth is that traffic forecasting models are probably as deficient as are evaluations of the costs per unit of environmental damage (e.g. Pickrell, 1989; UK House of Commons Committee of Public Accounts, 1988; Button *et al.*, 1982).

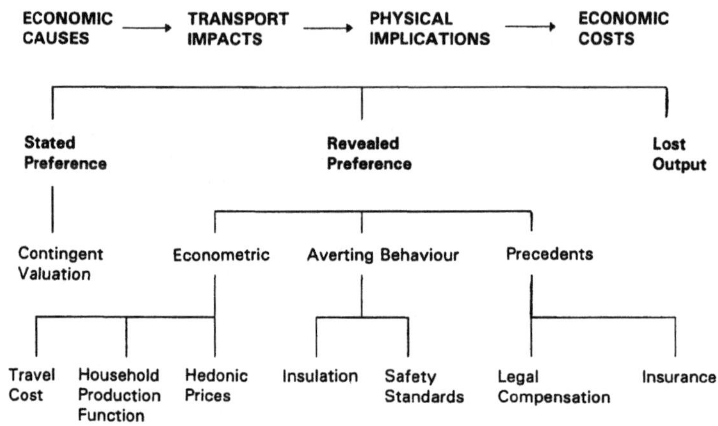

Figure 14.1. Methods of evaluating environmental damage.

The poor performance of forecasts, even after allowance is made for the potentially large amount of political capture that is evident (e.g. Kain, 1992), is surprising given the extremely large amount of intellectual effort expended in the field (especially relating to elasticities of demand). Surveys are to be found in Goodwin (1992) and Oum *et al.* (1992). As with the argument which follows regarding the usefulness of meta-analysis regarding the clarification of why environmental values can differ, so there is a case for deploying similar procedures to look at traffic forecasting models. Quite clearly, however, the difficulties go beyond traffic forecasting problems. They extend to the relating of these to ultimate economic and social outcomes. At the broadest level, the general empirical evidence from existing work does not even generate a consensus as to the impact of, say, infrastructure investments on the level of economic activity and location patterns. For a discussion of some of the key issues and the resultant problems of modelling and quantifying links see Rietveld (1989, 1994); Vickerman (1991) and Gwilliam (1979). A similar level of diversity is to be found amongst studies of a more macro type which have sought to discover the importance of infrastructure investment expenditures on national economic development – a survey of the US literature in this field is contained in Munnell (1992).

The alternative, and generally preferred, way of approaching the question of the social and environmental impacts of transport policy is through the direct effects on the costs of transport. Once again, however, even if the changes in physical traffic volumes and flows could be forecast accurately, there remain problems in trying to convert these into economic costs and benefits and then to express these in monetary units. The difficulty, though, is not so much a dearth of empirical evidence on the values to attach, but rather it is often due to an embarrassing richness of work. While we subsequently offer brief summaries of travel time (Table 14.1), noise nuisance (Table 14.3) and land use effects (Table 14.5), one could equally produce listings of findings regarding the values associated with local environmental degradation (e.g. Quinet, 1990; Verhoef, 1994; see also Button, 1993), accidents (a useful introductory reference to the literature in this area is Jones-Lee, 1990), and other relevant costs and benefits.

14.3 Illustrating Meta-analysis in Transport Evaluation

Given the number of studies that have surveyed the numerous and diverse implications of transport investments, it is somewhat surprising that few attempts have been made to analyse the results systematically. As indicated earlier, the surveys that have been done have relied upon traditional literature review techniques. There is no pretence that a fully rigorous meta-analysis is being conducted, but rather some simple results are offered looking at studies in three specific areas of transport impact evaluation: the value of travel time (relevant for congestion costs) (Section 14.3.2), noise nuisance (Section 14.3.3) and land use impacts (relevant when considering the pecuniary distribution of external impacts of transport) (Section 14.3.4).

The studies selected for the meta-analysis are based upon secondary, albeit very up-to-date and extensive, sources rather than representing the author's own selection. This has its defects and the study arrays could well be extended but, equally, it does have the advantage of removing the biases of the author from the study selection process. A range of meta-analysis techniques is available and selection depends upon the nature of the issue under consideration and the way the studies reviewed offer up their results. The estimation procedures adopted here are certainly not of the most technical kind but are based upon regressions of the variables of interest on a set of variables that encapsulate some of the key differentiating features of each study. This contrasts, for instance, to the work in labour absence and turnover which follows Hunter *et al.* (1982) and looks at simple correlations. This approach is preferred, because it does allow both a range of possible effects to be examined simultaneously and discrete factors to be incorporated.

14.3.1 Travel Time Savings

For a very simple illustration of how a basic meta-analysis may offer new insights use can be made of the survey of travel time values provided Waters (1992) which are set out in Table 14.1. This study throws up a substantial range of values for travel time as a percentage of the wage rate (the coefficient of variation being 0.652). While not capable of embracing a comprehensive set of the categories of variables set out in Equation 1, nor indeed to include a full set of items within each variable vector, a simple linear regression does provide guidance as to the way the meta-analysis can be applied to explore the reasons underlying this dispersion.

Table 14.1. A sample of estimated values of travel time savings.

Study	*Country*	*Value of time as % of wage rate*	*Trip purpose*	*Mode*
Beesley (1965)	UK	33-50%	Commuting	Auto
Quarmby (1967)	UK	20-25%	Commuting	Auto, Transit
Stopher (1968)	UK	21-32%	Commuting	Auto, Transit
Oort (1969)	USA	33%	Commuting	Auto
Thomas & Thompson (1970)	USA	86%	Interurban	Auto
Lee & Dalvi (1971)	UK	30%	Commuting	Bus
		40%	Commuting	Auto
Wabe (1971)	UK	43%	Commuting	Auto, Subway
Talvitte (1972)	USA	12-14%	Commuting	Auto, Transit
Hensher & Hotchkiss (1974)	Australia	2.70%	Commuting	Hydrofoil, Ferry
Kraft & Kraft (1974)	USA	38%	Interurban	Bus
McDonold (1975)	USA	45-78%	Commuting	Auto, Transit
Ghosh *et al.* (1975)	UK	73%	Interurban	Auto
Guttman (1975)	USA	63%	Leisure	Auto
		145%	Commuting	Auto
Hensher (1977)	Australia	39%	Commuting	Auto
		35%	Leisure	Auto
Nelson (1977)	USA	33%	Commuting	Auto
Hauer & Greenough (1982)	Canada	67-101%	Commuting	Subway
Edmonds (1983)	Japan	42-49%	Commuting	Auto, Bus, Rail
Deacon & Sinstelie (1985)	USA	52-254%	Leisure	Auto
Troung and Hensher (1985)	Australia	105%	Commuting	Auto, Transit
Guttman & Menashe (1986)	Israel	59%	Commuting	Auto, Bus
Fowkes (1986)	UK	27-59%	Commuting	Rail, Coach
Hau (1986)	USA	46%	Commuting	Auto, Bus
Chui & McFarland (1987)	USA	82%	Interurban	Auto
Mohring *et al.* (1987)	Singapore	60-129%	Commuting	Bus
Cole & Sherman (1990)	Canada	93-170%	Commuting	Auto
		116-165%	Leisure	Auto

Sources: As listed in Waters (1992). An up-dating of this paper is Waters (1993).

The general model tested is VT = *f* (C, M, T) (14.1)

where:

- VT is the value of non-work travel time found across the various studies expressed as a percentage of income;
- C is a vector of country dummy variables with the UK as the base;
- M is a vector of mode dummy variables with auto travel as the base;
- T is a vector of trip purpose dummies with commuting as the base.

The results and summary statistics are set out in Table 14.2. Where there is a range provided by any study, the mid-point is selected for the estimation process. No allowance is made for the standard deviations which surround the reported values. The estimation is obviously rather crude, and ideally one would perhaps like to embrace a wider range of explanatory variables in the analysis, not least of which would be an indicator of the statistical methods they each employed, but nevertheless this very elementary exercise throws up some interesting results. Of particular importance is the fact that none of the independent variables that are often seen as explaining variations in the value of travel time emerge as statistically significant at any reasonable confidence level (there is an extensive literature on the valuation of travel time; for useful summaries we refer to Waters (1992), Hensher (1989) and Sharp (1981)). In effect there is relatively robust evidence that non-work time is valued well above 50% of the wage rate. This is also a somewhat higher figure than used by the UK Department of Transport in its standard evaluation programme (UK Department of Transport, 1989a).

Table 14.2. Coefficients from value of time model.

	Coefficients	*s.e.*
Intercept	58.369	18.414
North America	26.287	17.805
Australia	-12.894	25.587
Japan	10.586	42.229
Israel	24.586	42.229
Singapore	60.586	42.229
Multimodal studies	0.878	23.548
Ferries	27.923	25.422
Interurban trips	-23.955	17.437
Leisure trips	-42.475	48.499

$R^2 = 0.396$; N = 29; * significant at 95% level

Looking at the coefficients, and appreciating their statistical limitations, there are indications that the value of time savings related to both inter-urban and leisure travel is less than for urban trips that would conform to *a priori* expectations if the commuting travel time is less predictable because of congestion. International differences are also shown up in the results, perhaps indicative of variations in national tastes and priorities.

14.3.2 Traffic Noise

Traffic noise nuisance is regularly cited in urban studies as one of the major disadvantages of city life (Button, 1993). It has been studied extensively, often using hedonic price indices but more recently contingent valuation methods, but there are several problems in placing a value on it. Some of these problems are more to do with the technical measurement of noise and the links between traffic and noise levels rather than with the difficulties of evaluation *per se*.

In particular, noise is contextual; an acceptable noise level in a recreation facility such as a sports stadium is unacceptable in the home. Added to this 'grey' noise (the background noise we generally take for granted) is growing and this makes determination of the overall costs of noise more difficult. There are also measurement problems concerned with weighting such things as the level, pitch, duration, frequency and so on with the result that there are numerous different physical scales identifying the impacts of traffic noise. Most studies have, therefore, attempted to express total traffic noise as a percentage of GDP and it is Verhoef's (1994) survey of this work that we examine here (see Table 14.3). The subsequent regressions only make use of those studies which are based on primary calculations and excludes those which extract estimates from surveys of other works.

Table 14.3. Survey of findings relating to traffic noise.

Authors	*Country*	*Percentage of GDP*	*Method*
Ringheim (1983)	Norway	0.22	Loss in property value
		0.17	Sleep loss
		0.07	Existing protection
		0.12	Potential vehicle protection programme
Wick (1987)	Germany	0.15	Productivity loss
		1.45	Lower property values
IRT (1983)	France	0.30-2.27	Outside insulation
Lambert (1987)	France	0.04	Loss in property values
UIC (1987)	Netherlands	0.02	Government expenditure on abatement
Opschoor (1987)	Netherlands	0.02	Falling prperty values
Sharp *et al.* (1976)	UK	11.18	Reducing traffic noise
Quinet (1989)	General	0.01	Comparison of studies
Kanafani (1983)	Europe	0.1-0.2	Comparison of studies
	USA	0.06-0.12	Comparison of studies
Bouladon (1979)	General	0.3-1.0	Abatement at source
Bleijenberg (1988)	Netherlands	0.03-0.08	Extra prevention and property value decline
Van der Meijs (1988)	Netherlands	0.03-0.08	Insulation
		0.12-0.24	Abatement at source
Dogs *et al.* (1991)	Germany	0.03	Roads - avoidance costs
		0.03	Rail - avoidance costs
		0.52	Roads - WTP
		0.22	Rail - WTP
Grupp (1986)	Germany	0.04-0.10	Avoidance costs
		0.02-0.05	Falling house prices

Source: As listed in Verhoef (1994).

As seen in the table, there are considerable variations both in the results generated by the individual studies (the coefficient of variation is 1.189) and in the approaches employed to arrive at these results. Pearce and Markandya (1989) present a survey of results looking at the values for noise generated in studies conducted in the USA, Canada and Switzerland using hedonic price techniques. They also find significant variations between studies when expressing the impact of one unit change in L_{eq} on house prices.

The set of independent variables employed reflects the estimation procedure used in each study (which, since several studies employed combined procedures, does not need the exclusion of a base variable to meet identification requirements) and the country studied. A variety of other factors, such as the definition of what exactly is meant by traffic noise nuisance in each study, would ideally have been included but information on this is not readily available. One finds (Table 14.4) that while much of the variation is explained by the variables included, many are not significant. Variations according to technique indicate that there is a tendency for willingness-to-pay and production loss techniques to generate higher valuations.

Table 14.4. Coefficients of noise value model.

	Coefficients	*s.e.*
Intercept	-0.324	0.666
Property value	0.223	0.598
Production loss	1.701	0.636
Protection	0.614	0.636
WTP	1.064	0.837
Source abatement	0.467	0.934
Norway	-2.461	1.646
France	0.675	0.467
Netherlands	0.037	0.414

$R^2 = 0.706$; N = 13; * significant at 95% level

14.3.3 Land Use Impacts

Transport investments are often seen as influential on land use patterns and infrastructure enhancement has been a consistent feature of many countries' regional policy. While these positive effects may be captured by looking at the direct transport costs and benefits, an alternative is to examine the implications for the local land or labour markets. There is little consensus, however, amongst academics as to whether transport has any beneficial effects on land use or whether it can be detrimental (Button *et al.*, 1995). Table 14.5, for example, summarises the results of a number of recent studies in this area and the divergent findings can clearly be seen from the qualitative comments relating to each.

This leads to one of the difficulties of applying regression analysis in this context, namely that the effect of interest in transport infrastructure is not immediately quantified. What can be done, however, to gain some insight into the influences involved is to simply treat the development effect as discrete – namely whether it was negative or positive (strictly, since the dependent variable is constrained a transformation is needed to prevent predicted values falling outside of the designated range. This is not attempted here, but because the model is not being used for forecasting purposes this should not pose a major problem). Multiple least squares regression analysis is then undertaken relating variations in this dichotomous variable to: the type of transport investment being studied (namely whether it was a road based investment or otherwise); the area to which the study relates (whether it was North America or elsewhere) and the nature of the estimation methodology used (whether it was survey based or not).

Table 14.5. Summary of findings of studies examining transport and economic changes.

Author/technique	*Geographical site*	*Infrastructure*	*Conclusions*
Botham (1980) (1)	28 Zones (UK)	Changing nature of highways	Small centralising effect on employment
Briggs (1981) (2)	Non-metropolitan counties (US)	Provision of highways	Interstate highways do not guarantee of county development
Bruinsma *et al.* (1993) (3)	Urban area ((Netherlands)	Orbital motorway	No changes to office prices
Cleary & Thomas (1973) (3)	Regional level (UK)	New estuarial crossing	Little relocation but changes in firm's operations
Dasgupta & Webster (1992) (4)	Leeds, Bilbao & Dortmund	Inner and outer ringroads	No clear centralisation effect
Dodgson (1974) (1)	Zones in North (UK)	New motorway	Small effect on employment
Eagle *et al.* (1987) (5)	87 Counties (US)	New highway expenditure	No increase in employment
Evers *et al.* (1987) (6)	Regional level (Netherlands)	High-speed rail	Some effect on employment
Forrest *et al.* (1987) (7)	Metropolitan areas (US)	Light rapid transit	Property blight - good for urban renewal
Guiliano (1986) (8)	Metropolitan areas (US)	Urban beltways	Small but significant effects on regional development patterns
Judge (1983) (9)	Regional level (UK)	New motorway	Small economic impact
Langley (1981) (10)	Highway corridor (US)	Highway	Devaluated property in area
Mackie *et al.* (1986) (3)	Regional level (UK)	New estuarial crossing	Small overall effect
Mills (1981) (1)	Metropolitan areas (US)	Interstate highways	No significant effect on location patterns
Moon (1986) (11)	Metropolitan areas (US)	Highway interchanges	Existence of interchange villages
Pickett (1984) (4)	Local districts (UK)	Light rapid transit	Properties close to the line benefit
Stephandes (1980) (1)	87 Counties (US)	New highway expenditure	Could affect employment - depends on county's economy
Stephandes *et al.* (1986) (1)	87 Counties (US)	New highway expenditure	Some positive association with employment
Twark *et al.* (1986) (1)	Non-urban areas (US)	Highway interchanges	Some growth forecast
Watterson (1986) (1)	Metropolitan areas (UK)	Light rapid transit	Modest growth in land use
Webster *et al.* (1988) (4)	Cities in Japan, UK, Germany & Spain	Orbital motorways	Small decentralising effect
Wilson *et al.* (1982) (3)	Regional level (Canada)	Existing highways	Transport affects location decisions but not development

Notes on techniques employed:

1. Regression techniques
2. Path analytical procedures
3. Interviews and surveys
4. Land use-transportation models
5. Time-series analysis
6. Multi-sectoral potential
7. Hedonic price indices
8. Survey of studies
9. Traffic flows
10. House resales
11. Aerial photography

Sources: As listed in Button *et al.* (1995).

Table 14.6 lists the coefficients estimated from the regressions. The fit is hardly inspiring and individual coefficients are often not significant. The general indication from this work is that the method of analysis employed in these studies seems to exert an influence on results produced and that survey based approaches produce a lower probability of a positive link between transport and an increase in economic activity than do other procedures. Equally, there is some indication that roads are less influential in obtaining a positive land use impact than are other modes.

Table 14.6. Coefficients of economic impacts model.

	Coefficients	*s.e.*
Intercept	1.108	0.208
Roads	-0.329	0.222
North American	-0.260	0.212
Survey	-0.694	0.265

R^2 = 0.324; N = 22; * significant at 95% level

14.4 Conclusions

The focus of this paper has been on questioning whether when attempting to forecast the economic implications of a major transport investment or policy shifts we make the best use of the information that is available to us. The need to evaluate these economic implications is certainly not new and a considerable body of knowledge concerning, for instance, the magnitude of key parameters and the strength of various forecasting techniques, has now been built up. What one finds in practice, however, is that this is often ignored or the accumulation of the results of previous work is treated in a non-rigorous fashion with a proclivity for 'eye-balling' rather than statistically analyzed. Put another way, a considerable amount of resources are often expended on individual studies and the most sophisticated econometric procedures are deployed but when it comes to bringing together our accumulated knowledge about key issues or parameters there is a tendency to rely exclusively on judgement and feel, rather than on statistical analysis.

Forecasting the overall impact of new pieces of transport infrastructure or policy reform is never easy, and to forecast the various implications that such changes have on the various different affected parties is even more problematic. The difficulties are compounded by the various interested parties who in their efforts at rent seeking attempt to capture the forecasting process. The use of meta-analysis, while not entirely removing this element of subjectivity, does offer the scope for reducing it and, where it does remain, the prospect of making it much more transparent. Equally, however, as demonstrated in the almost back-of-the-envelope calculations offered in this paper, the task of assembling the data base and then statistically analysing it is far from straightforward.

PART D

CONCLUSIONS AND GUIDELINES

CHAPTER 15 SUMMARY CONCLUSIONS AND GUIDELINES

15.1 Merits of Meta-analysis

Policy-makers are increasingly confronted with various and often seemingly diverse results from a large number of studies of environmental phenomena. Much the same holds for analysts working on more technical aspects of environmental issues. In preparing studies, the technical analyst can draw upon extensive bodies of existing work undertaken elsewhere that might contain useful information or insights. When the characteristics and results of previous studies are broadly similar, especially in terms of the problems considered and the approaches used, one question is whether this material can be systematically processed to generate comprehensive and concise conclusions that may be transferable to other areas of work. If this is so, then rather than performing an additional new in-depth study, it may be possible to elicit relevant information by looking at earlier studies dealing with similar questions.

Dependent on the type of issue, the methods involved and the variety of cases, it may be possible to adopt a well-defined framework for summarising, assessing, comparing, averaging, evaluating or apprehending common elements in other studies. This variety of processes has been the subject of this volume. Appropriate application of meta-analysis can help to improve our understanding of environmental issues. It can help make better use of prior information and knowledge. It can help remove some of the subjectivity from analysis and forecasting, or at least make judgments more transparent. Additionally, it can lead to greater clarity as to where future efforts in environmental economic analysis can most gainfully be deployed. Finally, it may offer insights into phenomena for which no specific study has yet been conducted.

15.2 Challenges of Meta-analysis in Environmental Economics

Traditionally, meta-analysis procedures employ statistical analysis techniques to help remove some of the subjectivity inherent in other reviewing procedures, such as literary reviews. In many fields where, unlike most situations in economics, studies are characterised by mutual consistency in terms of, for example, data, scope, type of questions, performance indicators, statistical methods and presentation, it is generally relatively easy to synthesise results of separate studies. The major problem in these cases, especially when dealing with new medical treatments or diseases when comparing psychological experiments, is often the practical one of bringing together the information but, even here, new computerised data and information bases are making things easier.

Given the relative lack of meta-analysis studies in environmental economics, one issue to consider is what kinds of policy or technical studies can be assessed. For a strict, statistics-based meta-analysis suitable studies must ideally satisfy several conditions. They should have been performed in a standardised way, for instance, with respect to their set-up and the presentation of results. This means that the findings should not have been presented selectively in order to use them primarily as vehicles of advocacy rather than education. Studies should also involve quantitative factors and identical units, or at least results which can be reduced to some common unit or index.

The statistical techniques used in traditional meta-analyses are diverse. This is in part because of the differing nature of the topics addressed and the range of statistical issues that emerge when looking at heterogeneous, quantitative works. The formal statistical techniques developed for traditional meta-analysis, however, can seldom be directly applied to studies in which quasi-scientific methods are used. Four types of constraints are evident when trying to apply statistical meta-analysis to economic studies:

- In the medical field, where meta-analysis has most widely been applied, there is considerably more standardisation in the way results are reported than in economics, and this difference is particularly pronounced in newer research fields such as environmental economics.
- There is a tendency in economics only for positive results to be published making it difficult to incorporate quite legitimate negative results in any overview. This limitation may even be more relevant for environmental economics, because this new discipline has relatively few journals for dissemination of research results.
- There are numerous consultancy studies undertaken in economics, by both the public and private sectors, that are not accessible for strict statistics-based meta-analysis. The result is the inevitable exclusion of a body of valuable information. In addition, problems of confidentiality are likely to arise in competitive markets, where information can confer market power, as well as in cases where release may lead to public concern. Even if these studies are made public, findings are often not presented objectively but rather as vehicles of policy advocacy.
- In terms of need for similarity in output measures, there are clearly areas where methodological problems exist in comparing economic impacts. This tends to be a particular problem when qualitative factors are involved or diverse units of measurement are used. For environmental economic topics this can be specifically relevant, since it often involves the linking of a range of physical-environmental to monetary-economic aspects.

In case studies where strict experiments are used to generate data for estimation of models and for the testing of hypotheses, the exogenous variables are more explicitly controlled than in studies where data comes from coincidental, actual processes. The first typifies work in the natural sciences and the latter that in social sciences.

Application of traditional statistical meta-analysis to case studies in the fields of environmental and ecological economics is even more intricate. Reliable and useful environmental, ecological and economic data are often scarce. This is, in part, because the interaction between economic activities and environmental degradation has only been intensively studied for a relatively short time, and systematic observation and data collection has started only recently. Furthermore, there is a fundamental problem in that many environmental effects are often delayed or do not occur in simple cause-effect loops but are embedded in a complex network of ecological, hydrological and physical processes. Additionally, there is only limited experience with implementing policies and management scenarios at various administrative levels, so that little systematic and long-standing information about environmental policy effects is available.

The upshot is that it can be useful to apply meta-analysis methods to some categories of environmental research and policy assessment. Directly transferring meta-analytical techniques from sciences such as medicine or psychology to environmental economics is, however, generally problematic. We have therefore started from a general meta-analytical framework in designing and applying various types of formal and less formal meta-analytical tools and techniques. In addition, attention has been given to the actual performing of meta-analysis of individual study results.

15.3 Step-wise Procedures for Meta-analysis

Cooper (1982) has presented a five-stage model of the integrative review as a research process. We mainly follow the Cooper model. At each stage, he lists the question asked, primary function, procedural variations and associated threats to validity (see Table 15.1). In other words, description of the functions, procedures, and threats to scientific validity engendered by each step in the reviewing process are included in his model. It should be noted that Cooper's is not the only framework. Rosenthal (1991), for example, makes a classification in three stages that can be followed when undertaking a meta-analysis: (1) defing research results; (2) retrieving and assessing research results and (3) comparing and combining research results. Further, in a more recent study Pettiti (1994) makes the following classification in stages: (1) information retrieval; (2) data collection and (3) data-analysis.

There is, thus, some diversity in using stage approaches. The model of research synthesis presented in Table 15.1 is neutral with regard to the purpose of review (Cooper, 1988). The analysis and case studies presented in earlier chapters suggest that this type of framework has considerable merit. This model will, therefore, be used as framework for the remainder of this section because its integrative qualities make it suitable for the reviewing process which is necessary for meta-analysis.

15.3.1 Formulation of the Problem

The presentation of a case study used in a meta-analysis should ideally include the exact formulation of the objective or objectives of the study. Objectives of meta-analysis can be formulated as one of the following:

- to summarize for a set of studies what the overall relationship is between two or more variables that have been investigated in each study;

Table 15.1. The integrative review as a research project.

Stage characteristics	*Stage of research*			
	Problem formulation	Data collection	Data retrieval and evaluation	Analysis and interpretation
Research question asked	What evidence should be included in the review?	What procedures should be used to find relevant evidence?	What retrieved evidence should be included in the review?	What procedures should be used to make inferences about the literature as a whole?
Primary function in review	Constructing definitions that distinguish relevant from irrelevant studies	Determining which sources of potentially relevant studies to examine	Apply criteria to separate 'valid' from 'invalid' studies	Synthesizing valid retrieved studies
Procedural differences that create variation in review conclusions	1. Differences in included operational definition 2. Differences in operational detail	Differences in the research contained in sources of information	1. Differences in quality criteria 2. Differences in the influence of non-quality criteria	Differences in rules of inference
Sources of potential invalidity in review conclusions	Narrow concepts might make review conclusions less definitive and robust	1. Accessed studies might be qualitatively different from the target population of studies. 2. People sampled in accessible studies might be different from target population of people	1. Non-quality factors might cause improper weighting of study information 2. Omissions in study reports might make conclusions unreliable	1. Rules for distinguishing patterns from noise might be inappropriate 2. Review-based evidence might be used to infer causality

Source: Adapted from Cooper (1982).

- to determine the factors that are associated with variations in the magnitude of relationships between two or more variables;
- to aggregate or average data that are correlated with each other or with other characteristics of the study to test hypotheses or to suggest hypotheses to be tested in subsequent specifically designed studies;
- to test whether one method gives significantly different results compared with other methods;
- to meta-predict investigated phenomena for cases not yet examined;
- to determine whether recent studies support the conclusions of prior ones;
- to examine several different theoretical models in order to further develop policy-relevant theories and practical strategies;
- to identify research issues and problems that must be addressed before a comprehensive synthesis can be performed.

Meta-analysis has been extremely powerful in case of sharply posed research questions. But in many cases systems are complex and this may need a modification of an existing theory and a verification via existing empirical research. This can require flexibility in research methods and subjects.

15.3.2 Collection of Data

Meta-analysis is concerned with commonalities across studies studies. In specific cases, however, one has to make sure that the outcome measurements are comparable. The sort of data available will influence the evidence from the various background studies. Gaps in knowledge are, however, often a relevant concern if the set of studies under review is biased. In general, it is important to use both primary and secondary sources. Primary sources embrace archival periodical literature, including dissertations and theses, while secondary sources include reviews, that cite and organize material found in primary sources and incorporate review periodicals and various abstract and citation archives.

The need for a balanced reviewing process also means that attention should be given to searching techniques. These techniques should complement one another to ameliorate systematic bias in the methods and results of the case studies. To guarantee an acceptable objective collection of study-material, several steps are important: a search of the literature using key words, an examination of lists of references in key research reports and a contacting of experts on the area concerned.

15.3.3 Data Retrieval and Evaluation

Meta-analysis should offer a mapping that is as complete as possible, otherwise file drawer problems may cause bias (Greenwald, 1975). Additionally, the time span covered by studies under consideration can result in bias (Cook, 1992). On the other hand, ideally one wants maximum uniformity. Given the conditional nature of relevant judgements, it is not possible to present a general set of guidelines for deciding whether or not a particular study should be retained in any meta-analysis, although efforts have been made in the past to do this (e.g., Bryant and Wortman, 1984). As a general rule one should include studies that, focus on the same phenomenon; use the same outcome measure; use the same population and use the same 'good' research design. Another important issue concerning data retrieval is that there are often clear cases of deficient or inferior data reporting. The following strategies can be used in this context:

- External sources can be used to obtain information about instrumentation that was not reported in the primary study;
- Additional data or clarification of procedures can be obtained from the original investigator;
- Deficiencies in reporting can be accounted for as part of the coding process, e.g. by explicitly recording the recorder's confidence in the codes assigned to extracted data.

The characteristics of studies often considered important in conducting meta-analyses may be either substantive or methodological. Substantive characteristics are those specific to the problem studied. The methodological characteristics of the study are more general. Glass *et al.* (1984) argue that measurement of study characteristics can be evaluated in terms of both its validity and reliability. In such cases, contextual, methodological and output aspects play a critical role. A basic list of important issues to be addressed in this case includes:

- Study context (variables regarding country of study, authors discipline, authors affiliation, source of research, type of publication and year of publication)
- Method (experimental groups, sample size, sampling, initial equivalence of experimental groups procedures for assignment to groups, attrition from experimental groups, characteristics of the control condition)
- Treatment/policy measures (characteristics of subjects, amount or intensity of policy exposure, characteristics of the policy condition)
- Outcome (descriptive outcome, statistical outcome, statistical information (effect size/outcomes).

15.3.4 Analysis and Interpretation

The findings of studies employed in a meta-analysis should be expressed in some common scale. Study findings can be expressed directly in cases where a simple statistical expression of results is common to all. Methods also exist for more complex cases where scales and methods of reporting findings vary between studies. In these cases the separate data points should be syntesized into a unified statement about the research problem.

Meta-analysis aims to extract common lessons from a set of similar studies. Often, however, a large spectrum of conclusions can be extracted, dependent on the aim of the experiment. Conclusions from meta-analyses could consist of: discovery of meaningful patterns; conceptualization of between-studies variation, factors explaining differences; making studies comparable and combinable; covariates (on the study-level and the subject-level); and predictions based on meta-regression models.

15.4 Techniques for Meta-analysis

A basic analytical problem in a meta-analysis is the choice of a proper technique. There exists a variety of techniques, more or less formalized or standardised. The most important of these are the traditional reviewing technique, statistically-based meta-analysis, meta-multicriteria analysis, epistemological analysis and the newly developed rough set analysis.

The use of quantitative techniques and statistics can, at least in part, reduce the disadvantages associated with the use of traditional approaches of a literary type. To use these techniques, however, it is necessary for the results of the individual studies to be complete and of a considerable homogeneity, a requirement which is rarely fulfilled in environmental economics, apart perhaps from certain areas within economic valuation of changes in environmental quality. In the choice of the technique, it is, therefore, always necessary to bear in mind the principle that it is not possible to generate information that the results of the individual studies do not contain.

In the context of meta-techniques, we have in particular stressed the importance of rough set theory. The aim of the rough set approach is to pinpoint possible regularities in all the information available, in order to draw from these elements and relationships which are not immediately evident. More exactly, this approach attempts to discover possible cause-effect relationships, to underline the importance and the strategic role of some data and the irrelevance of other data. Moreover, this approach does not require any preliminary or further information about the data available. It can easily approach qualitative data. It is also able to take into consideration any existing inconsistent information, which is not automatically eliminated *a priori*, because it is possible to obtain facts from both positive and negative examples.

With reference to decision-making, rough sets can be used to advantage in problems of choice, sorting and ranking of actions in the field of multi-attribute decision support. Moreover, this approach, in building a model of the decision-maker's preferences, does not require the specification of any parameters (nor, consequently, the complex information correlated to these parameters), but simply uses examples or previous experiences supplied by the decision-maker from which to obtain decisional rules that, discussed and acknowledged by the decision-maker, can then be used to aid the decision process.

An interesting comparison between statistics and rough set approach pertains to the role of the user. Once the model in statistics is fixed, there is a very limited possibility of intervention for the user. On the contrary, the rough set approach requires a constant participation of the user in all the stages of the analysis. In the first stage, it is important to model the imprecision of the data by choosing among different sets of thresholds. Afterwards, the user can choose among the different reducts those that better represent the essential features of the phenomenon. Then, when some rules are determined to represent the phenomenon, the user can judge about the real significance of the results of rough set analysis by a process of backtracking to the objects. Finally when the decision rules are matched with other objects in order to make some predictions it is important to compare the new objects with the original ones from which the considered decision rules have been derived.

The limitations of rough set analysis must, however, also be recognized. Unlike regression approaches it can result in lost information since continuous variables must be translated into discrete classes. If, for example, one has information or actual estimates from a series of case studies, these must be classified into groups to conduct a rough set analysis. Further, and linked to this, the adoption of combined continuous and dummy variable specifications allows for the direct inclusion of quantitative and many forms of quality variables in a meta-regression model. The merits of using the alternative approaches must, therefore, be weighed up with care. Furthermore, as also shown in some of the applications in Part C of the book, applying both methods for specific meta-analyses may render complementary or even mutually supportive information.

15.5 Epilogue

Empirical research is generally expensive, both financially and in terms of the manpower involved. Many fields of research in the social sciences are gradually building up a record of both empirical findings and of experiences in using alternative analytical techniques. There is also a growing number of studies looking at the implications of alternative policy approaches to socio-economic issues. This study has focused on suggesting ways in which, rather than simply continuing to add to the number of individual research projects being conducted, we can consolidate the experiences gained from previous work to both gain fresh insights into various issues and to clarify those fields where additional research might yield the greatest return. Environmental economics has been taken as the catalyst of our argument, because it is an extremely important area in its own right.

It is possible, by adopting a variety of statistical and other, 'softer' procedures, to extend and develop meta-analysis in ways which make it a powerful tool to supplement conventional, primarily literary, methods of reviewing existing bodies of work in environmental economics. This extends beyond the more conventional types of application found in the natural sciences, paralleled by point estimates of e.g. elasticities and externality valuations in economics, to the consideration of policy options and policy instruments.

Forecasting the overall impacts of environmental policies is never easy. The difficulties are compounded by the various interested parties who in their efforts at rent seeking attempt to capture the forecasting process. The use of meta-analysis, while not entirely removing this element of subjectivity, does offer the scope for reducing it and, where it does remain, the prospect of making it much more transparent.

INDEX

REFERENCES

Aaltink, A. (1992), *Economische Gevolgen van de Beperking van het Bestrijdingsmiddelengebruik in de Akkerbouw*, M.A. Thesis, Agricultural University, Wageningen.

Abelson, P.W. (1979), Property values and the value of amenities, *Journal of Environmental Economics and Management*, vol.6, 11-28.

Adams, J. (1993), The Emperor's old clothes: the curious comeback of cost-benefit analysis, *Environmental Values*, vol.2, 247-60.

Amemiya, T. (1981), Qualitative response models: a survey, *Journal of Economic Literature*, vol.19, 1483-536.

Anderson, R.J. and D.E. Wise (1977), The effects of highway noise and accessibility on residential property values, DOT-FH-11-8841, National Technical Information Service, Springfield, Va.

Antle, J. and P. Pingali (1994), Pesticides, productivity, and farmer health: a Philippine case study, *American Journal of Agricultural Economics*, vol.76, 418-430.

Archer, B.H. (1976), The anatomy of a multiplier, *Regional Studies*, vol.10, 71-77.

Archer, B.H. (1977), *Tourism in the Bahamas and Bermuda: two Case-studies*, Bangor occasional papers in economics, no.10, Cardiff.

Armstrong, H. and J. Taylor (1993), *Regional Economics and Policy*, Philip Allen, Oxford.

Arrow, K.J., R. Solow, P. Portney, E. Leamer, R. Radner and H. Schuman (1993), *Report of NOAA Panel on Contingent Valuation*, 58 Fed. Reg. 46.

Ayres, R.U. (1989), Industrial metabolism, in Ausubel, J.H. and H.E. Sladovich (eds.), *Technology and Environment*, National Academy Press, Washington D.C., 23-49.

Baaijens, S.R. (1996), *Raming van Toeristische Inkomens-multipliers met Meta-analyse*, Master's Thesis, Free University, Amsterdam.

Bailey, M.J., R.F. Muth and H.O. Nourse (1963), A regression method for real estate price index construction, *Journal of the American Statistical Association*, vol.58, 933-942.

Barde, J-P. and D.W. Pearce (eds.) (1991), *Valuing the Environment: Six Case Studies*, Earthscan, London.

Bartik, T.J. (1991), *Who Benefits from State and Local Economic Development Policies?*, W.E. Upjohm Institute, Kalamazoo.

Bennett, R.J. and R.J. Chorley (1978), *Environmental Systems*, Methuen & Co, London.

Bergh, J.C.J.M. van den (1996), *Ecological Economics and Sustainable Development: Theory, Methods and Applications*, Edward Elgar, Aldershot.

Bergh, J.C.J.M. van den, K.J. Button and S.M. Jongma (1995a), A meta-analytical framework for environmental economics, TRACE Discussion Paper TI 95-82, Tinbergen Institute, Amsterdam.

Bergh, J.C.J.M. van den, B. Matarazzo and G. Munda (1995b), Measurement and uncertainty issues in environmental management: a survey, TRACE Discussion Paper TI 95-81, Tibergen Institute, Amsterdam.

Blaug, M. (1992), *The Methodology of Economics: or How Economists Explain*, 2nd ed Cambridge University Press, Cambridge.

Böjo, J., K.G. Mäler and L. Unemo (1990), *Environment and Development: An Economic Approach*, Kluwer, Dordrecht.

Boulding, K.E. and S.B. Lundstedt (1988), Value concepts and justifications, in Peterson, G.L., B.L. Driver and R. Gregory (eds.), *Amenity Resource Valuation: Integrating Economics with Other Disciplines*, Venture Publishing, State College.

Bryant, F. and P.M. Wortman (1984), Methodological issues in meta-analysis of quasi-experiments, *New Directions in Program Evaluation*, vol.24, 25-42.

Button, K.J. (1992), *Market and Government Failures in Environmental Management: The Case of Transport*, OECD, Paris.

Button, K.J. (1993), *Transport, the Environment and Economic Policy*, Edward Elgar, Aldershot.

Button, K.J. (1994), Evaluation of transport externalities: what can we learn using meta-analysis?, Paper presented to the Advanced Studies Institute on Transport, Environment and Traffic Safety, Tinbergen Institute and Department of Spatial Economics, Free University, Amsterdam.

Button, K.J. (1995), Evaluation of transport externalities: what can we learn using meta-analysis?, *Regional Studies*, vol.29, 507-517.

Button, K.J. and J. Kerr (1996), Effectiveness of traffic restraint policies: a simple meta-regression analysis, *International Journal of Transport Economics*, vol.23, 213-225.

Button, K.J. and P. Rietveld (1996), Infrastructure investment and aggregate productivity - what does meta-analysis tell us?, Paper presented to the 5th World Congress of the Regional Science Association International, Tokyo.

Button, K.J. and T.G. Weyman-Jones (1992), Ownership structure, institutional organisation and measured X-efficiency, *American Economic Review, Papers and Proceedings*, vol.82, 439-445.

Button, K.J. and T.G. Weyman-Jones (1994), X-efficiency and technical efficiency, *Public Choice*, vol.80, 83-104.

Button, K.J., A.D. Pearman and A.S. Fowkes (1982), *Car Ownership Modelling and Forecasting*, Gower, Aldershot.

Button, K.J., S. Leitham, R.W. McQuaid and J.D. Nelson (1995), Transport and industrial and commercial location, *Annals of Regional Science*, vol.29, 189-206.

Button, K.J., W. Weidenfeld, G.W. Heinze, H.H. Kill, R. Münch, N. Walter, G. Topmann, H. Holzapfel and K.O. Schallaböch (1994), *Transport Policy – Ways into Europe's Future*, Bartelsmann Foundation, Gütersloh.

Bycic, P. (1992), Job performance and absenteeism: a review and meta-analysis, *Human Relations*, vol.45, 193-20.

Cameron, M. (1991), Transportation efficiency: tackling Southern California's air pollution and congestion, Environmental Defence Fund, Regional Institute of Southern California.

Carson, R.T., N.E. Flores, K.M. Martin and J.L. Wright (1996), Contingent valuation and revealed preference methodologies: comparing the estimates for quasi-public goods, *Land Economics*, vol.72, 80-99.

Chambers, G. and E. Lichtenberg (1994), Simple econometrics of pesticide productivity, *American Journal of Agricultural Economics*, vol.76, 407-417.

Chankong, V. and Y.Y. Haimes (1983), *Multiobjective Decision Making: Theory and Methodology*, North Holland, Amsterdam.

Cheslow, M.D. (1978), A preliminary analysis of a road pricing and transit improvement programme in Berkeley, Paper no.5050-3-6, Urban Institute, Berkeley.

Chow, C.G. (1975), *Analysis and Control of Dynamic Economic Systems*, Wiley, New York.

Chow, C.G. (1983), *Econometrics*, McGraw Hill, New York.

Coccossis, H. and P. Nijkamp (eds.) (1995), *Sustainable Tourism Development*, Avebury, Aldershot.

Collins, A. and A. Evans (1994), Aircraft noise and residential property values. An artificial neural network approach, *Journal of Transport Economics and Policy*, vol.28, 175-197.

Commission of the European Communities (1988), *Proposal for a Council Regulation for an Action Programme in the Field of Transport Infrastructure with a View to the Completion of an Integrated Transport Market in 1992*, Document COM(88) 340 Final, CEC, Brussels.

Commission of the European Communities (1992), *Transport Infrastructure*, Document COM(92) 231 Final, CEC, Brussels.

Commission on the Third London Airport (1972), *Report*, HMSO, London.

Cook, T.D. (1992), *Meta-analysis for Explanation: A Casebook*, Russell Sage Foundation, New York.

Cook, T.D. and L.C. Leviton (1980), Reviewing the literature: a comparison of traditional methods with meta-analysis, *Journal of Personality*, vol.48, 449-72.

Cooper, H.M. (1982), Scientific guidelines for conducting integrative research reviews, *Review of Educational Research*, vol.52, 291-302.

Cooper, H.M. (1988), Organising knowledge synthesis: a taxonomy of literature reviews, *Knowledge in Society*, vol.8, 104-126.

Cooper, H. and L.V. Hedges (eds.) (1994), *The Handbook of Research Synthesis*, Russell Sage Foundation, New York.

Cracknell, J.A., B.V. Martin, A.D. May and D. Pickering (1975), Physical restraint: Greater London's cordon restraint study, *Traffic Engineering and Control*, vol.16, 368-372.

Cropper, M.L. and W.E. Oates (1992), Environmental economics: a survey, *Journal of Economic Literature*, vol.30, 675-740.

DeGroot, M.H. (1970), *Optimal Statistical Decisions*, McGraw-Hill, London.

Diamond, P.A. (1996), Testing the internal consistency of contingent valuation surveys, *Journal of Environmental Economics and Management* (forthcoming).

Drakopoulos, J. (1994), Probabilities, Possibilities and Fuzzy Sets, Stanford University, Working Paper of the Knowledge Systems Laboratory, Stanford.

Dubgaard, A. (1991), Pesticide regulation in Denmark, in Hanley, N. (ed.), *Farming and the Countryside: an Analysis of External Costs and Benefits*, CAB International, Wallingford, 48-58.

Dubois D. and H. Prade (1980), *Fuzzy Sets and Systems: Theory and Applications*, Academic Press, New York and London.

Dygert, P.K. (1973), *Estimation of the cost of aircraft noise to residential activities*, Ph.D. dissertation, University of Michigan, Michigan.

Economist (1991), Under the metascope, vol.319, no.7707, 93-94.

Elhorst, J. (1990), *De Inkomensvorming en de Inkomensverdeling in de Nederlandse Landbouw Verklaard vanuit de Huishoudproduktietheorie*, Landbouw-Economisch Instituut, The Hague.

Elliot, W. (1975), The Los Angeles affliction: suggestions for a cure, *Public Interest*, vol.38, 119-128.

Elmberg, C.M. (1973), The Gothenburg traffic restraint scheme, *Transportation*, vol.1, 1-27.

Emerson, F.C. (1969), *The determinants of residential value with special reference to the effects of aircraft nuisance and other environmental features*, Ph.D. dissertation, University of Minesota, Minesota.

Farrell, D. and C.L. Stamm (1988), Meta-analysis of the correlates of employee absence, *Human Relations*, vol.41, 211-227.

Fiset, L., P. Milgrom and J. Tarnai (1994), Dentists response to financial incentives in a mail order survey of malpractice liability experience, *Journal of Public Health Dentistry*, vol.54, 68-72.

Fiske, D.W. (1983), The meta-analytic revolution in outcome research, *Journal of Consulting Psychology*, vol.51, 65-70.

Fletcher, J. (1989), Input-output analysis and tourism impact studies, *Annals of Tourism Research*, vol.16, 514-519.

Foster, C.D. (1968), Future research needs of transport planners, in Young, M. (ed.), *Forecasting and the Social Sciences*, Heinemann, London.

Foster, C.D. and M.E. Beesley (1963), Estimating the social benefits of constructing an underground railway in London, *Journal of the Royal Statistical Society*, vol.126A, 46-56.

Fowkes, A.S., D.S. Milne, C.A. Nash and A.D. May (1993), The distributional impact of various road charging schemes for London, Working Paper 400, Institute for Transport Studies, University of Leeds, Leeds.

Freeman III, A.M. (1993), *The Measurement of Environmental and Resource Values: Theory and Methods*, Resources for the Future, Baltimore.

Funtowicz, S.O. and J.R. Ravetz (1987), Qualified quantities - towards an arithmetic of real experience, in Forge, J. (ed.), *Measurement, Realism and Objectivity*, Dordrecht/Reidel.

Funtowicz, S.O. and J.R. Ravetz (1990), *Uncertainty and Quality in Science for Policy*, Kluwer, Dordrecht.

Gamble, H.B. (1974), *The Influence of Highway Environmental Effects on Residential Property Values*, Institute for Research on Land and Water Resources, University Park, Pa.

Gautrin, J-F. (1975), An evaluation of the impact of aircraft noise on property values with a simple model of urban land rent, *Land Economics*, vol.51, 80-86.

Ghali, M.O. and M.J. Smith (1994), Designing time-of-day signal plans which reduce urban traffic congestion, *Traffic Engineering and Control*, vol.35, 555-560.

Giaoutzi, M. and P. Nijkamp (1993), *Decision Support Models for Regional Sustainable Development*, Avebury, Aldershot.

Gilbert, J.P., R.J. Light and F. Mosteller (1975), Assessing social innovations: an empirical base for policy, in Bennet, C.A. and A.A. Lumsdaine (eds.), *Evaluation and Experiment*, Academic Press, New York, 39-193.

Glass, G.V. (1976), Primary, secondary, and meta-analysis of research, *Educational Research*, vol.5, 3-8.

Glass, G., B. MacGaw and M.L. Smith (1984), *Meta-analysis in Social Research*, Sage, Beverly Hills.

Gomez-Ibanez, J.A. and G.R. Fauth (1980), Downtown auto restraint policies; the costs and the benefits for Boston, *Journal of Transport Economics and Policy*, vol.14, 133-153.

Goodwin, P.B. (1992), A review of the new demand elasticities with special reference to short and long run effects of price changes, *Journal of Transport Economics and Policy*, vol.26, 155-70.

Greco, S., B. Matarazzo and R. Slowinski (1995), Rough set approach to multi-attribute choice and ranking problems, ICS Research Report 38/95, Warsaw University of Technology, Warsaw.

Greco, S., S. Lo Cascio and B. Matarazzo (1996), Rough set approach to stock selection: an application to the Italian market, in Bertocchi, M., E. Cavalli and S. Komlosi (eds.), *Modelling Techniques for Financial Markets and Bank Management*, Physica Verlag, Heidelberg, 192-211.

Greenwald, A.G. (1975), Consequences of prejudice against the null hypothesis, *Psychological Bulletin*, vol.82, 1-20.

Grzymala-Busse, J.W. (1991), *Managing Uncertainty in Expert Systems*, Kluwer, Dordrecht.

Grzymala-Busse, J.W. (1992), LERS - A system for learning from examples based on rough sets, in Slowinski, R. (ed.), *Intelligent Decision Support Handbook of Applications and Advances of the Rough Set Theory*, Kluwer, Dordrecht.

Guzzo, N., S.E. Jackson and R.A. Katzell (1987), Meta-analysis analysis, *Research in Organisational Behaviour*, vol.9, 407-422.

Gwilliam, K.M. (1979), Transport infrastructure investments and regional developments, in Bowers, J.K. (ed.), *Inflation, Development and Integration – Essays in Honour of A.J. Brown*, University Press, Leeds, 241-262.

Hackett, R.D. (1990), Age, tenure and employment absenteeism, *Human Relations*, vol.43, 601-619.

Hackett, R.D. and R.M. Guion (1985), A reevaluation of the absenteeism-job satisfaction relationship, *Organisational Behaviour and Human Decision Processes*, vol.35, 340-381.

Hall, F.L., B.E. Breston and S.M. Taylor (1978), *Effects of highway noise on residential property values*, Transportation Research Record 686, National Academy of Sciences, Washington, 38-43.

Hartmut, T. and T. Pharaoh (1994), Car-free city centres, *Transportation*, vol.21, 231-247.

Hau, T.D. (1990), Developments in transport policy: electronic road pricing, *Journal of Transport Economics and Policy*, vol.24, 203-214.

Hauser, G. and R. Maggi (1993), Missing transport networks: a policy perspective, in Banister, D. and J. Berechman (eds.), *Transport in a Unified Europe - Policies and Challenges*, North-Holland, Amsterdam, 173-193.

Hausman, J.A. (ed.) (1993), *Contingent Valuation: A Critical Assessment*, North-Holland, Amsterdam.

Hedges, L.V. and I. Olkin (1985), *Statistical Methods for Meta-analysis*, Academic Press, New York.

Hensher, D.A. (1989), Behavioural and resource values of travel time savings: a bicentennial update, *Australian Road Research*, vol.19, 223-29.

Hersh, H.M., A. Caramazza, and H.H. Brownell (1979), Effects of context on fuzzy membership functions, in Gupta M.M., Regade R.K. and Yager R.R. (eds.), *Advances in Fuzzy Set Theory and Applications*, Amsterdam, North Holland, 389-408.

Hettelingh, J.-P. (1989), *Uncertainty in Modeling Regional Environmental Systems*, Ph.D. dissertation, Free University, Amsterdam.

Hey, J.D. (1991), *Experiments in Economics*, Blackwell, Oxford.

Hirschleifer, J. and J.G. Riley (1979), The analytics of uncertainty and information - an expository survey, *Journal of Economic Literature*, vol.17, 1375-1421.

Hoevenagel, R. (1994), *The Contingent Valuation Method: Scope and Validity*, Ph.D. dissertation, Free University, Amsterdam.

Houwen, C. van der (1993), *Economische Modellering van Akkerbouw-bedrijven in Zuidelijk Flevoland in Relatie tot Milieu en EG-beleid*, Ministerie van Verkeer en Waterstaat, Lelystad.

Hunter, J.E. and F.L. Schmidt (1990), Dichotomization of continuous variables: the implications for meta-analysis, *Journal of Applied Psychology*, vol.75, 334-349.

Hunter, J.E., F.L. Schmidt and G. Jackson (1982), *Advanced Meta-analysis: Quantitative Methods for Cumulating Research Findings A Cross Studies*, Sage, Beverly Hills.

Janssen, R. (1992), *Multi-objective Decision Support for Environmental Management*, Kluwer, Dordrecht.

Jarrell, S.B. and T.D. Stanley (1990), A meta-analysis of the union-nonunion wage gap, *Industrial and Labor Relations Review*, 54-67.

Johansson, P-O. (1987), *The Economic Theory and Measurement of Environmental Benefits*, Cambridge University Press, Cambridge.

Johansson, P-O. (1995), *Evaluating Health Risks: An Economic Approach*, Cambridge University Press, Cambridge.

Jones-Lee, M.W. (1990), The value of traffic safety, *Oxford Reviews of Economic Policies*, vol.6, 39-60.

Jong, W. de (1991), *Milieu-economische Modellering: Inkomens- en Milieu-effecten van een Heffing op Bestrijdingsmiddelen in de Nederlandse Akkerbouw*, M.A. Thesis, Agricultural University, Wageningen.

Kain, J.F. (1992), The use of straw men in the economic evaluation of rail transport projects, *American Economic Review, Papers and Proceedings*, vol.82, 487–493.

Kandelaars, P., J.B. Opschoor and J.C.J.M. van den Bergh (1995), A dynamic simulation model for materials-product chains: An application to gutters, TRACE Discussion Paper TI 95-61, Tinbergen Institute, Amsterdam.

Keeney, R.L. and H. Raiffa (1976), *Decisions with Multiple Objectives: Preferences and Value Trade-Offs*, Wiley, New York.

Kelsey, D. (1988), The economics of chaos or the chaos of economics, *Oxford Economic Papers*, vol.40, 1-31.

Kemeny, J.G. (1959), *A Philosopher Looks at Science*, Van Nostrand Reinhold, New York.

Khan, H., C.F. Seng and W.K. Cheon (1990), Tourism multiplier effects on Singapore, *Annals of Tourism Research*, vol.17, 408-418.

Kissinger, H.A. (1960), *The Necessity for Choice: Prospects of American Foreign Policy*, Harper and Row, New York.

Liu, J., T. Var and A. Timur (1984), Tourist-income multipliers for Turkey, *Tourism Management*, December 1984, 280-287.

Koslowksy, M. and A. Sagie (1993), On the efficacy of credibility intervals as indicators of moderator effects in meta-analytic research, *Journal of Organizational Behavior*, vol.14, 695-699.

Kungliga Skogs- och Lantbruksakademien (1989), *Minskad Bekaempning i Jordbruket*, Sweden.

Langley, J.C. (1976), Adverse impacts of the Washington beltway on residential property values, *Land Economics*, vol.52, 54-65.

Larsen, O.I. (1988), The toll ring in Bergen, Norway: the first year of operation, *Traffic Engineering and Control*, vol.16, 216-222.

Lenarcik, A. and Z. Piasta (1992), Discretization of condition attribute space, in Slowinski, R. (ed.), *Intelligent Decision Support*, Kluwer, Dordrecht.

Lenarcik, A. and Z. Piasta (1993), Probabilistic approach to decision algorithm generation in the case of continuous condition attributes, *Foundation of Computing and Decision Science*, vol.18, 213-223.

Lenarcik, A. and Z. Piasta (1994), Rough classifiers, in Ziarko, W.P. (ed.), *Rough Sets, Fuzzy Sets and Knowledge Discovery*, Springer-Verlag, Berlin.

Leung Y. (1988), *Spatial Analysis and Planning Under Imprecision*, North Holland, Amsterdam.
Levesque, T.J. (1994), Modelling the effects of airport noise on residential housing markets: a case study of Winnipeg International Airport, *Journal of Transport Economics and Policy*, vol.28, 199-210.
Light, R.J. (ed.) (1983), *Evaluation Studies Review Annual*, Sage, Beverly Hills.
Light, R.J. and D.B. Pillemer (1984), *Summing Up - The Science of Reviewing Research*, Harvard University Press, Cambridge MA.
Lowry, P.E. (1994), The structured interview - an alternative to the assessment-center, *Public Personnel Management*, vol.23, 201-215.
Lush, J.L. (1931), Predicting gains in feeder cattle and pigs, *Journal of Agricultural Research*, vol.42, 853-881.
Macan, T.H. (1994), Time management - test of a process model, *Journal of Applied Psychology*, vol.79, 381-391.
Mann, C. (1990), Meta-analysis in the breech, *Science*, vol.249, 476-480.
Maser, S.M., W.H. Riker and R.N. Rosett (1977), The effects of zoning and externalities on the price of land: an empirical analysis of Monroe county, New York, *Journal of Law and Economics*, vol.20, 111-132.
May, A.D. (1975), Supplementary licensing: an evaluation, *Traffic Engineering and Control*, vol.16, 162-167.
May, A.D. (1992), Road pricing: an international perspective, *Transportation*, vol.19, 313-333.
May, A.D. and C.A. Nash (1990), Assessing the Benefits and Incidence of Road Pricing in London, Institute for Transport Studies, University of Leeds, Leeds.
McMillan, M.L., B.G. Reid and D.W. Gillen (1978), An approach towards improved estimates of willingness to pay for public goods from hedonic price functions: A case of aircraft noise, Unpublished paper, University of Alberta, Alberta, Alberta.
McMillan, M.L., B.G. Reid and D.W. Gillen (1980), An extension of the hedonic approach for estimating the value of quiet, *Land Economics*, vol.56, 315-328.
Miezkowski, P. and A.M. Saper (1978), An estimate of the effects of airport noise on property values, *Journal of Urban Economics*, vol.5, 425-440.
Miller, G. and C. Everett (1982), Raising commuter parking prices - an empirical study, *Transportation*, vol.11, 105-129.
Milne, S.S. (1987), Differential multipliers, *Annals of Tourism Research*, no.4, 499-515.
Milne, S.S. (1992), Tourism and development in South Pacific microstates, *Annals of Tourism Research*, vol.19, 191-212.
Ministerie van Verkeer en Waterstaat (1988), *Road Pricing Project: Programme of Requirements*, Ministerie van Verkeer en Waterstaat, The Hague.
Miser, H.J. (1993), A foundation of concept of science appropriate for validation in operational research, *European Journal of Operational Research*, vol.66, 204-215.
Mishan, E.J. (1988), *Cost-benefit Analysis*, 3rd edition, Allen and Unwin, London.

Mitra, A., G.D. Jenkins and N.A. Gupta (1992), A meta-analytic review of the relationship between absence and turnover, *Journal of Applied Psychology*, vol.6, 879–889.

Mrozek (1985), Information systems and control algorithms, *Bulletin of the Polish Academy of Sciences, Series Technical Sciences*, vol.33, 195-212.

Mrozek (1989), Rough sets and dependency analysis among attributes in computer implementations of expert's inference models, *International Journal of Man-Machine Studies*, vol.30, 457-471.

Munda, G. (1993), *Fuzzy Information in Multicriteria Environmental Evaluation Models*, Ph.D. dissertation, Free University, Amsterdam.

Munda, G. (1995), *Multicriteria Evaluation in a Fuzzy Environment. Theory and Applications in Ecological Economics*, Physica-Verlag, Berlin.

Munda, G., P. Nijkamp and P. Rietveld (1993), Information precision and evaluation methods, in Giardina, E. and A. Williams (eds.), *Efficiency in the Public Sector*, Edward Elgar, Aldershot, 161-183.

Munda, G., P. Nijkamp and P. Rietveld (1994), Qualitative multicriteria evaluation for environmental management, *Ecological Economics*, vol.10, 97-112.

Munnell, A.H. (1992), Infrastructure investment and economic growth, *Journal of Economic Perspectives*, vol.6, 189–98.

MVA Consultancy (1995), The London Congestion Charging Research Programme, MVA Consultancy, London.

Nelson, J.P. (1978), *Economic Analysis of Transportation Noise Abatement*, Ballinger, Cambridge, Mass.

Nelson, J.P. (1980), Airports and property values: a survey of recent evidence, *Journal of Transport Economics and Policy*, vol.19, 37-52.

Neumann, J. von, and O. Morgenstern (1944), *Theory of Games and Economic Behavior*, Princeton University Press, Princeton.

Nijkamp, P. and E. Blaas (1994), *Impact Assessment and Evaluation in Transportation Planning*, Kluwer, Boston.

Nijkamp, P., H. Leitner, H. and N. Wrigley (eds.) (1984), *Measuring the Unmeasurable*, Kluwer, Dordrecht.

Nijkamp, P., P. Rietveld, and H. Voogd (1990), *Multicriteria Evaluation in Physical Planning*, North-Holland, Amsterdam.

Norwich A.M. and I.B. Turksen (1982), The fundamental measurement of fuzziness, in Yager R.R. (ed.), *Fuzzy Set and Possibility Theory*, New York, Pergamon.

O'Bryne, P.H., J.P. Nelson and J.J. Seneca (1985), Housing values, census estimates, disequilibrium, and the environmental cost of airport noise: A case study of Atlanta, *Journal of Environmental Economics and Management*, vol.12, 169-178.

Ones, D.S., M.K. Mount, M.R. Barrick and J. Hunter (1994), Personality and job performance: a critique of the Tett, Jackson, and Rothstein (1991) meta-analysis, *Personnel Psychology*, vol.47, 147-156.

Opschoor, J.B. (ed.) (1992), *Environment, Economy and Sustainable Development*, Wolters-Noordhoff, Amsterdam.

Organisation for Economic Cooperation and Development (1992), *Market and Government Failures in Environmental Policy: The Case of Transport*, OECD, Paris.
Orth, L.C. and R.P. Martin (1994), Interactive effects of student temperament and instruction method on classroom-behaviour and achievement, *Journal of School Psychology*, vol.32, 149-166.
Oskam, A. (1992), *Economic Analysis of the Use of Fertilizer and Pesticides in Dutch Arable Farming*, Agricultural University, Wageningen.
Oskam, A., H. van Zeijts, G. Thijssen, G. Wossink and R. Vijftigschild (1992), *Pesticide Use and Pesticide Policy in the Netherlands*, Agricultural University, Wageningen.
Oswald, A.J. (1991), Progress and microeconomic data, *Economic Journal*, vol.101, 75-80.
Oum, T.H., W.G. Waters and J-S Yong (1992), Concepts of price elasticities of transport demand and recent empirical estimates, *Journal of Transport Economics and Policy*, vol.27, 139-154.
Paik, I.K. (1972), *Measurement of environmental externality in particular reference to noise*, Ph.D. dissertation, Georgetown University, Georgetown.
Pawlak, Z. (1982), Rough sets, *International Journal of Information and Computer Sciences*, vol.11, 341-356.
Pawlak, Z. (1991), *Rough Sets. Theoretical Aspects of Reasoning About Data*, Kluwer, Dordrecht.
Pawlak, Z. and R. Slowinski (1993), Decision analysis using rough sets, *International Transactions in Operations Research*, vol.1.
Pearce, D.W. (1989), *Tourism Development*, Longman, London.
Pearce, D.W. (1992), Green economics, *Environmental Values*, vol.1, 3-13.
Pearce, D.W., and A. Markandya (1989), *Environmental Policy Benefits: Monetary Valuation*, OECD, Paris.
Pearce, D.W. and C.A. Nash (1989), *The Social Appraisal of Projects*, Macmillan, London.
Pearce, D.W. and P.K. Turner (1990), *Economics of Natural Resources and the Environment*, Harvester Wheatsheat, New York.
Pepping, G. and M. de Bruijn (1991), *Tourism: Emphasis on Selective Development*, M.A. Thesis, Vrije Universiteit, Amsterdam.
Petterson, O., J. Petterson, V. Johansson, H. Fogelfors, R. Sigvald and B. Johnsson (1989), *Minskad Bekämpning i Jordbruket; Möjligheter och Konsekvenser*, Kungliga Skogs- och Lantbruksakademiens Tidskrift, no.128, 379-396.
Pettiti, D.B. (1994), *Meta-analysis, Decicion Analysis, and Cost-effectiveness Analysis: Methods for Quantitative Synthesis in Medicine*, Oxford University Press, New York.
Phillips, J.M. (1994), Farmer education and farmer efficiency: a meta-analysis, *Economic Development and Cultural Change*, vol.43, 149-165.
Phillips, J.M. and E.P. Goss (1995), The effect of state and local taxes on economic development: a meta-analysis, *Southern Economic Journal*, vol.62, 320-333.

Pickrell, D.H. (1989), *Urban Rail Transit Projects: Forecast versus Actual Ridership and Costs*, US Department of Transportation, Cambridge MA.

Pleeter, S. (1980), *Economic Impact Analysis*, Martinus Nijhoff, Boston.

Plott, C.R. (1982), Industrial organisation theory and experimental economics, *Journal of Economic Literature*, vol.20, 1484-1527.

Ploudre, G. and D. Yeung (1989), A model of industrial pollution in a stochastic environment, *Journal of Environmental Management and Economics*, vol.16, 97-105.

Price, I. (1974), *The social cost of airport noise as measured by rental changes: The case of Logan Airport*, Ph.D. dissertation, Boston University, Boston.

Quinet, E. (1990), *The Social Costs of Land Transport*, Environmental Monograph 32, OECD, Paris.

Rathbone, D.B. (1992), Telecommuting in the United States, *ITE Journal*, 40-45.

Ravetz, J.R. (1971), *Scientific Knowledge and its Social Problems,* Oxford University Press, Oxford.

Reus, J., H. Weckseler and G. Pak (1994), *Towards a Future EC Pesticide Policy*, Centre for Agriculture and Environment, Utrecht.

Ridker, R.G. (1967), *Economic Costs of Air Pollution: Studies in Measurement*, Praeger, New York.

Ridker, R.G. and J.A. Henning (1967), The determinants of residential property values with special reference to air pollution, *Review of Economics and Statistics*, vol.49, 246-257.

Rietveld, P. (1980), *Multiple Objective Decision Methods in Regional Planning*, North-Holland, Amsterdam.

Rietveld, P. (1989), Infrastructure and regional development: a survey of multiregional economic models, *Annals of Regional Science*, vol.23, 255-274.

Rietveld, P. (1994), Spatial economic impacts of transport infrastructure supply, *Transport Research,* vol.28A, 329–342.

Robertson, I.T. and A. Kinder (1993), Personality and job competences: the criterion-related validity of some personality variables, *Journal of Occupational and Organisational Psychology*, vol.66, 225-244.

Rosenthal, R. (1963), On the social psychology of the psychological experiments: the experimenter's hypothesis as unintended determinant of experimental results, *American Scientist*, vol.51, 268-283.

Rosenthal, R. (1991), *Meta-analytic Procedures for Social Research*, Sage, Beverly Hills.

Rust, R.T. and B. Cooil (1994), Reliability-measures for qualitative data - theory and implications, *Journal of Marketing Research*, vol.31, 1-14.

Schipper, Y.J.J. (1996), On the valuation of aircraft noise: a meta-analysis, Paper presented at the 36th congress of the European Regional Science Association, Zürich.

Schulte, J. (1983), *Der Einfluss eines Begrenzten Handelsduenger- und Pflanzenbehandlungsmitteleinsatzes auf Betriebsorganisation und Einkommen Verschiedener Betriebssysteme*, Ph.D. Dissertation, Rheinischen Friedrich-Wilhelms-Universität, Bonn.

Schwartz, J. (1994), Air pollution and daily mortality: a review and a meta analysis, *Environmental Economics*, vol.64, 36-52.

Scott, D.K. and G.S. Taylor (1985), An examination of conflicting findings on the relationship between job satisfaction and absenteeism: a meta-analysis, *Academy of Management Journal*, vol.28, 599-612.

Segerson, K. (1988), Uncertainty and incentives for nonpoint pollution control. *Journal of Environmental Economics and Management*, vol.15, 87-98.

Sharp, C. (1981), *The Economics of Time*, Routeledge, London.

Slowinski, R. (ed.) (1992a), *Intelligent Decision Support: Handbook of Applications and Advances of the Rough Sets Theory*, Kluwer, Dordrecht.

Slowinski, R. (1992b), A generalization of the indiscernibility relation for rough set analysis of quantitative information, *Rivista di Matematica per le Scienze Economiche e Sociali*, vol.15, 65-78.

Slowinski, R. (1993), Rough set learning of preferencial attitude for multi-criteria decision making, in Komorowski, J. and Z.W. Ras (eds.), *Methodologies for Intelligent Systems, Lecture Notes in Artificial Intelligence*, vol.689, Springer-Verlag, Berlin.

Slowinski, R. and J. Stefanowski (1992), RoughDAS and RoughClass software implementations of the rough sets approach, in Slowinski, R. (ed.), *Intelligent Decision Support: Handbook of Applications and Advances of the Rough Sets Theory*, Kluwer, Dordrecht.

Slowinski, R. and J. Stefanowski (1994), Handling various types of uncertainty in the Rough Set Approach, in Ziarko, W.P. (ed.), *Rough Sets, Fuzzy Sets and Knowledge Discovery*, Springer, Berlin.

Slowinski, R. and D. Vanderpooten (1995), Similarity relation as a basis for rough approximation, ICS Research Report 53/95, Warsaw University of Technology, Warsaw.

Smith, V.K. (1989), Can we measure the economic value of environmental amenities?, *Southern Economic Journal*, vol.56, 865-878.

Smith, V.K. and Y. Kaoru (1990a), Signals or noise: explaining the variation in recreation benefit estimates, *American Journal of Agricultural Economics*, vol.72, 419-433.

Smith, V.K. and Y. Kaoru (1990b), What have we learned since Hotelling's letter? A meta-analysis, *Economics Letters*, vol.32, 267-272.

Smith, V.K. and J. Chin Huang (1993), Hedonic models and air pollution: Twenty five years and counting, *Environmental and Resource Economics*, vol.3, 381-394.

Smith, V.K. and J. Chin Huang (1995) Can markets value air quality? A meta-analysis of hedonic property values models, *Journal of Political Economy*, vol.103, 209-227.

Smith, V.K. and L. Osborne (1996), Do contingent valuation estimates pass a "Scope" test? A meta-analysis, *Journal of Environmental Economics and Management*, vol.31, 287-301.

Speilberger, C.D. and E.C. Reheiser (1994), The job stress survey - measuring gender differences in occupational stress, *Journal of Social Behaviour and Personality*, vol.9, 199-218.

Spielburg, F. (1978), Transportation improvements in Madison, Wisconsin: preliminary analysis of pricing programs for roads and parking in conjunction with transit charges, Paper no.5050-3-7, Urban Institute, Washington.

Stanley, T.D. and S.B. Jarrell (1989), Meta-regression analysis: a quantitative method of literature surveys, *Journal of Economic Surveys*, vol.3, 161-170.

Steel, R.P. and N.K. Ovaille (1984), A review and meta-analysis of research on the relationship between behavioural intentions and employee turnover, *Journal of Applied Psychology*, vol.69, 673-686.

Stefanowski, J. (1992), Rough sets theory and discriminant methods as tools for analysis of information systems. A comparative study, *Foundations of Computing and Decision Sciences*, vol.17, 81-98.

Steuer, R.E. (1986), *Multiple Criteria Optimization: Theory, Computation and Application*, Wiley, New York.

Stewart, M.B. and K.F. Wallis (1981), *Introductory Econometrics*, Basil Blackwell, Oxford.

Storstadstrafik (1989), *Traffic in Big Cities*, Volume 3, Stockholm.

Sturtevant, L.A., F.R. Johnson and W.H. Desvousges (1995), A meta-analysis of recreational fishing, Triangle Economic Research, Durham, North Carolina.

Sujan, H., B.A. Weitz and N. Kumar (1994), Learning orientation, working smart and effective selling, *Journal of Marketing*, vol.58, 39-52.

Szymanski, D.M., S.G. Bharadwaj and P.R. Varadarajan (1993), An analysis of the market share-profitability relationship, *Journal of Marketing*, vol.57, 1-18.

Theil, H. (1964), *Optimal Decision Rules for Government and Industry*, North-Holland, Amsterdam.

Tinbergen, J. (1956), *Economic Policy: Principles and design*, North-Holland, Amsterdam.

Tu, P.N.V. and E.A. Wilman (1992), A generalized predator-prey model: uncertainty and management, *Journal of Environmental Economics and Management*, vol.23, 123-138.

UK Department of Transport (1989a), *COBA 9 Manual*, DOT, London.

UK Department of Transport (1989b), *National Traffic Forecasts (Great Britain)*, HMSO, London.

UK House of Commons Committee of Public Accounts (1988), *Road Planning*, HMSO, London.

Usategui, J.M. (1990), Uncertain irreversibility, information and transformation costs, *Journal of Environmental Economics and Management*, vol.19, 73-85.

Uyeno, D., S.W. Hamilton and A.J.G. Biggs (1993), Density of residential land use and the impact of airport noise, *Journal of Transport Economics and Policy*, vol.27, 3-18.
Var, T. and J. Quayson (1985), The multiplier impact of tourism in the Okanagan, *Annals of Tourism Research*, vol.12, 497-514.
Varian, H.R. (1992), *Microeconomic Analysis*, Norton, New York.
Vany, A.S. de (1974), The revealed value of time in air travel, *Review of Economics and Statistics*, vol.56, 77-82.
Vaughan, R.J. and L. Huckins (1975), *The Economics of Expressway Noise Pollution Abatement*, P-5475, The Rand Corporation, Santa Monica.
Verhoef, E. (1994), External effects and social costs of road transport, *Transportation Research*, vol.28A, 273–288.
Verhoeven, J., A. Wossink and J. Reus (1994), An environmental yardstick in farm economic modelling of future pesticide use: the case of arable farming, *Netherlands Journal of Agricultural Science*, vol.42, 331-341.
Vickerman, R.W. (1991), Transport infrastructure in the European Community: new developments, regional implications and evaluation, in Vickerman, R.W. (ed.), *Infrastructure and Regional Development*, Pion, London, 36-50.
Viscusi, W.K. (1993), The value of risks to life and health, *Journal of Economic Literature*, vol.31, 1912-1946.
Voogd, H. (1983), *Multicriteria Evaluation for Urban and Regional Planning*, Pion, London.
Wachter, K.W. (1988), Disturbed by meta-analysis, *Science*, vol.241, 1407-1408.
Wallsten, T.S., D.V. Budescu, A. Rapoport, R. Zwick and B. Forsyth (1986), Measuring the vague meaning of probability terms, *Journal of Experimental psychology: General*, vol.115, 348-365.
Walsh, R.G., D.M. Johnson and J.R. McKean (1989), Market values from two decades of research on recreational demand, in Link, A.N. and V.K. Smith (eds.), *Advances in Applied Economics Volume 5*, JAI Press, Greenwich.
Walters, C. (1986), *Adaptive Management of Renewable Resources*, MacMillan, New York.
Wang, M.C., G.D. Haertel and H.J. Walberg (1993), Toward a knowledge-base for school learning, *Review of Educational Research*, vol.63, 249-294.
Waters, W.G. (1992), Values of travel time savings used in road project evaluation: a cross-country/jurisdiction comparison, *Australian Transportation Research Forum*, Bureau of Transport and Communications Economics, Canberra.
Waters, W.G. (1993), Variations in the value of travel time savings: empirical studies and the values for road project evaluation, *mimeo.*
Waters, W.G. (1995), Values of travel time savings in road transport project evaluation, paper to the 7th World Conference of Transport Research, Sydney.
Weitzman, M.L. and D.L. Kruse (1990), Profit sharing and productivity, in Binder, A. (ed.), *Paying for Productivity*, Brooking Institution, Washington.

Williams, C.R. and L.P. Livingstone (1994), Another look at the relationship between performance and voluntary turnover, *Academy of Management Journal*, vol.37, 269-298.

Windle, C. (1994), The shift to research synthesis, *Administration and Policy in Mental Health*, vol.21, 263-267.

Winkler, R.L. and W.L. Hays (1975), *Statistics, Holt, Rinehart and Winston*, New York.

Winston, C. (1985), Conceptual developments in the economics of transportation: an interpretative survey, *Journal of Economic Literature*, vol.23, 57-94.

Wit, C.T. de (1992), Resource use efficiency in agriculture, *Agricultural Systems*, vol.40, 125-151.

Witt, L.A. and L.G. Nye (1992), Gender and the relationship between perceived fairness of pay or promotion and job satisfaction, *Journal of Applied Psychology*, vol.77, 910-917.

Wofford, J.C. and L.Z. Liska (1993), Path-goal theories of leadership: a meta-analysis, *Journal of Management*, vol.19, 857-876.

Wolf, F. (1986), *Meta-Analysis: Quantitative Methods for Research Synthesis*, Sage, Beverly Hills.

Woodside, A.G., Th.M. Beretich and M.A. Lauricella (1993), Meta-analysis of effect sizes based on direct marketing campaigns, *Journal of Direct Marketing*, vol.7, 19-33.

Wossink, G. (1993), *Analysis of Future Agricultural Change: a Farm Economics Approach Applied to Dutch Arable Farming*, Ph.D. Dissertation, Agricultural University, Wageningen.

Wossink, A. and J. Renkema (1994), Farm economic modelling of integrated plant production, in: Struik, P. et al., *Plant Production on the Threshold of a New Century*, Kluwer, Dordrecht.

Yu, P.L. (1985), *Multi Criteria Decision Making: Concepts, Techniques and Extensions*, Pelenum Press, New York.

Zadeh, L.A. (1965), Fuzzy sets, *Information and Control*, vol.8, 338-353.

Zeleny, M. (1982), *Multiple Criteria Decision Making*, McGraw Hill, New York.

Ziarko (1993), *Analysis of Uncertain Information in the Framework of Variable Precision Rough Sets*, Foundation of Computing and Decision Sciences, vol.18, 3-4.

Zimmermann, H.J. (1986), *Fuzzy Set Theory and Its Applications*, Kluwer-Nijhoff, Boston.

MIX
Papier aus verantwortungsvollen Quellen
Paper from responsible sources
FSC® C105338

If you have any concerns about our products,
you can contact us on
ProductSafety@springernature.com

In case Publisher is established outside the EU,
the EU authorized representative is:
Springer Nature Customer Service Center GmbH
Europaplatz 3, 69115 Heidelberg, Germany

Printed by Libri Plureos GmbH
in Hamburg, Germany